T0181334

Synthese Library

Studies in Epistemology, Logic, Methodology, and Philosophy of Science

Volume 435

The aim of *Synthese Library* is to provide a forum for the best current work in the methodology and philosophy of science and in epistemology. A wide variety of different approaches have traditionally been represented in the Library, and every effort is made to maintain this variety, not for its own sake, but because we believe that there are many fruitful and illuminating approaches to the philosophy of science and related disciplines.

Special attention is paid to methodological studies which illustrate the interplay of empirical and philosophical viewpoints and to contributions to the formal (logical, set-theoretical, mathematical, information-theoretical, decision-theoretical, etc.) methodology of empirical sciences. Likewise, the applications of logical methods to epistemology as well as philosophically and methodologically relevant studies in logic are strongly encouraged. The emphasis on logic will be tempered by interest in the psychological, historical, and sociological aspects of science.

Besides monographs *Synthese Library* publishes thematically unified anthologies and edited volumes with a well-defined topical focus inside the aim and scope of the book series. The contributions in the volumes are expected to be focused and structurally organized in accordance with the central theme(s), and should be tied together by an extensive editorial introduction or set of introductions if the volume is divided into parts. An extensive bibliography and index are mandatory.

More information about this series at http://www.springer.com/series/6607

María Laura Martínez Rodríguez

Texture in the Work of Ian Hacking

Michel Foucault as the Guiding Thread of Hacking's Thinking

María Laura Martínez Rodríguez
Department of History and Philosophy of Science,
Institute of Philosophy,
School of Humanities and Education Sciences
University of the Republic
Montevideo, Uruguay

ISSN 0166-6991 ISSN 2542-8292 (electronic)
Synthese Library
ISBN 978-3-030-64787-2 ISBN 978-3-030-64785-8 (eBook)
https://doi.org/10.1007/978-3-030-64785-8

This Springer imprint is published by the registered company Springer Nature Switzerland AG
The registered company address is: Gewerbestrasse 11, 6330 Cham, Switzerland

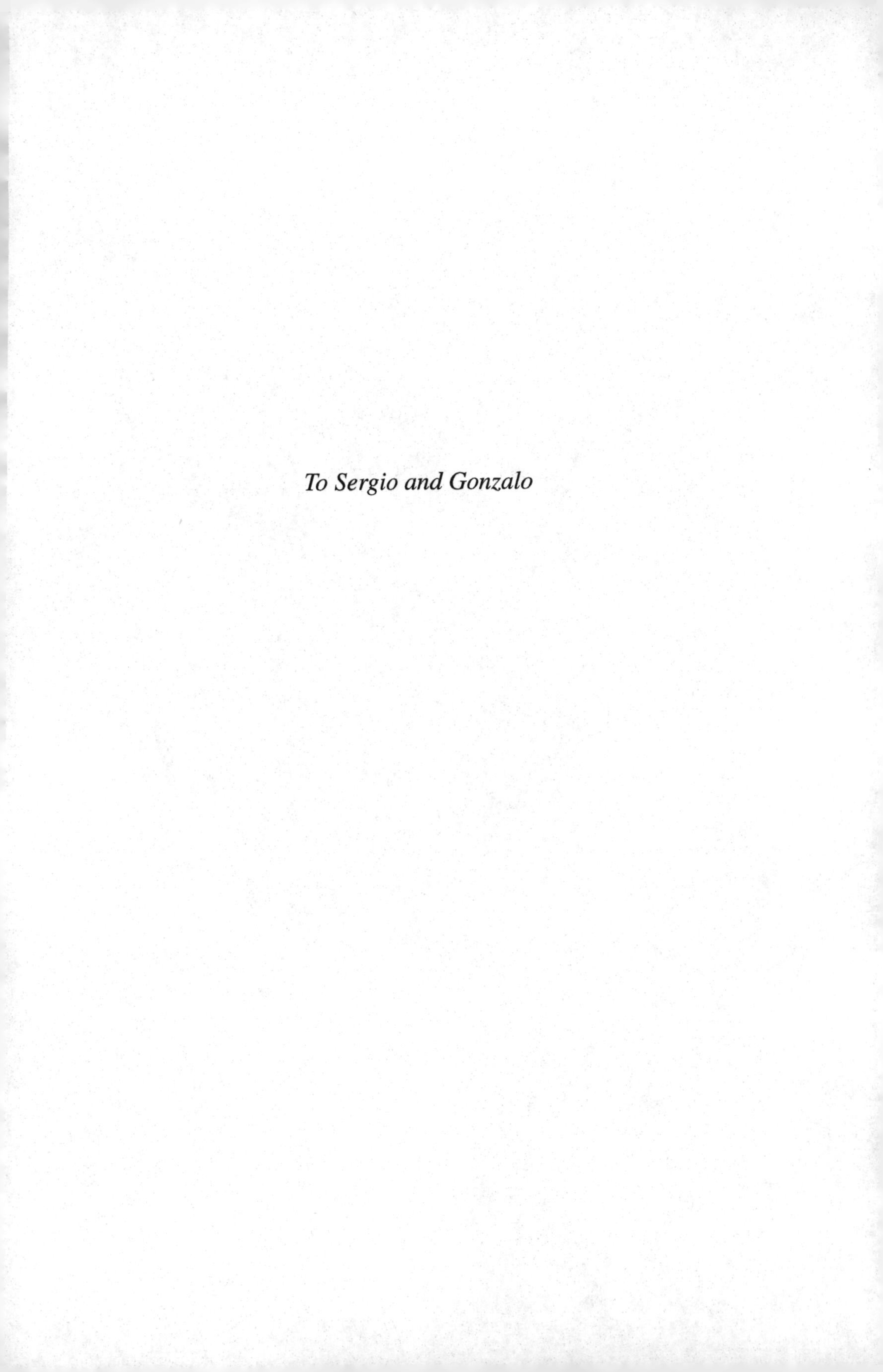

To Sergio and Gonzalo

Acknowledgments

This volume is a by-product of my doctoral and postdoctoral research, directed by Lucía Lewowicz and Otávio Bueno, respectively, whom I thank. My special thanks to Otávio for encouraging me to publish my research and his invaluable feedback as editor of Springer's Synthese Library during the publishing process.

My deepest gratitude to Prof. Mario Otero (in memoriam) and Prof. Alción Cheroni, my teachers and mentors in the Department of History and Philosophy of Science, Universidad de la República (Uruguay).

My thanks to Cecilia Molinari and Jared Hanson-Park for the careful translation and revision of this text, and to the anonymous reviewer whose feedback contributed to improving this book.

To Karina Silva, I thank her for her constant support.

I dedicate this book to my husband Sergio, for his love and patience during all these years and for his faith and encouragement in all my projects, and to my son Gonzalo, for so much love and so many kisses.

Introduction

> *The best three-word summary of my life, intellectual and other, is "I am curious." My curiosity gets me into a lot of trouble, not the least of which is that I follow up more different types of topics than I have time and energy to devote to them.*
>
> Ian Hacking (Vagelli 2014: 269)

Ian Hacking has defined himself as a philosopher trained in the analytic tradition (2002a:70–71). However, several of his works have an eminently historic mood, and he himself has acknowledged the profound influence philosophers of the so-called continental tradition have exerted on an important part of his work, in particular Michel Foucault. Hacking has defined *The Emergence of Probability* (1975a) as an archaeology strongly influenced by *Les mot et les choses* (1966) (*The Order of Things*); *The Taming of Chance* (1990a) as a Foucauldian genealogy, and in *Mad Travelers* (1998a) he remarks that his debt to Foucault is considerable and that in many of his books published after 1975 he has acknowledged the deep influence of that thinker on his work (Hacking 1998a:85–86).

Gianni Vattimo (2000:14) has remarked in the preface to D'Agostini's (2000) book that, in many senses, the separation or opposition between analytic and continental thought is perhaps the question that summarizes the characteristic problem of contemporary philosophy.

Whereas, according to D'Agostini, the image of these two great parallel lines can still be considered dominant; the divergence between both traditions has been enriched by articulations and distinctions, in such a way that the field has become increasingly complex, to the point that spaces of convergence have been outlined. However, the dialogue between both ways of doing philosophy is still difficult, which is reflected in the obstacles presented to those who have tried to join both positions. This is not surprising if we take into account that what is at stake are two kinds of philosophical practice and not merely two kinds of philosophical methods.

Hacking's space of convergence should not surprise us, given another influence that Hacking acknowledges, namely Ludwig Wittgenstein, whom he considers a philosopher that proposes a middle way between analytic and continental philosophers (Hacking 1975b: 176). However, Hacking formulates an interpretation of the problem that interchanges its terms, based on his idea that analytic philosophy is substantially poorer than its continental counterpart, and that the former has been overcome by the latter. In this sense it is worth noting that it is not only a question of Hacking being merely influenced by Foucault, but that the Canadian philosopher has the objective of bridging analytic and continental philosophy in such a way that neither loses its potential.

Hacking characterizes his way of doing philosophy as *taking a look* at the rich complexity of the world: "*It is not philosophy as conversation. It is philosophy as hard work. Or to use understatement, it is less talking than taking a look*" (2002a:71). In *Logic of Statistical Inference* (1965), he tries to understand statistical reasoning; *Why Does Language Matter to Philosophy?* (1975b) deals with philosophy of language, whereas *Representing and Intervening* (1983a) is a philosophical study on the role of experiment in science, where he aims to invert the traditional hierarchy of theory over experimentation. *Mad Travelers* deals with the case of an individual that suffers from hysterical fugue, using it as a framework for the discussion of recent "epidemics" of transient mental illnesses. *The Social Construction of What?* (1999a) deals with the phenomenon of the social construction of the social. The aforementioned works are representative—not exhaustive—examples of this diversity. However, Hacking argues that even though he has never felt a great need to unify his work, this is based on a network of connections some parts of which can be identified (Álvarez Rodríguez 2002: 8–9). The first one would be constituted by his works on probability; Michel Foucault is another part of the network, and Hacking mentions a third framework: the notion of making up people. These examples suggest the existence of a series of topics that make up a network such that several of his works can be considered to belong to more than one fraction. This underscores his particular way of working, as well as suggesting the existence of a thread across his work. One of the purposes of this book is to look for and define what I consider to be one of these threads: the influence of Michel Foucault's thought on Hacking's work.

Hacking has remarked that his debt to Foucault dates back to his first contact with the French philosopher around 1967–1968, when he read *Madness and Civilization: A History of Insanity in the Age of Reason* (1964),[1] even though *The Order of Things* was the book that changed his vision as an analytic philosopher. The first publication under this influence was his conference at the British Academy in 1973, "Leibniz and Descartes: Proof and Eternal Truths."

In general terms, the influence of Foucault is evident in his works on styles of scientific reasoning and the human sciences. Hacking noticed that Foucault's works could answer several scientific dilemmas that had existed for some time but became

[1] Abbreviated edition of *Histoire de la folie à l'âge classique* (1961).

evident after the publication of Thomas Kuhn's *The Structure of Scientific Revolutions* (1962). Under this influence, he proposed his notion of *style of scientific reasoning*, as well as a series of notions belonging to the human sciences such as *historical ontology, making up people, looping effect, and ecological niche,* among others. However, it is my impression that Foucault's influence extends to the rest of Hacking's work.

In this sense, the main idea of this work is that, despite the apparent diversity in Hacking's corpus, there is an underlying texture, woven by his interest in the analysis of the historical and situated conditions of possibility of the emergence of concepts and objects, and that this interest is due to Foucault's influence on his thought. In other words, I defend the influence of Michel Foucault's thought as a thread that runs across Ian Hacking's work.

Hacking has acknowledged that Michel Foucault constitutes one of the parts of this web. However, I believe that Foucault's thought, more than constituting a part of Hacking's work, is the texture that weaves together the structure; it appears more or less visibly, but it is always there. And it is so from the moment that Hacking was stimulated and interested in a Foucauldian analysis of the historical conditions of possibility for ideas, to the point that the French philosopher exemplifies what philosophy is for Hacking: "*a way of analyzing and coming to understand the conditions of possibility for ideas [...]*" (Hacking 1981a:76).

It is important to point out that, as this book shows, Hacking is not interested in Kantian transcendental conditions of possibility, and he goes even further than Foucault in this aspect. Hacking is interested in theoretical, historical, pragmatic, semantic, and social conditions of possibility. In his works, he analyzes how certain concepts and objects are possible and how they are historically altered.

This notion of historical and situated conditions of possibility, inherited from Foucault but at the same time different from his, impregnates, in my view, the whole of Hacking's work.

In Chap. 1, "Taking a Look at Ian Hacking's Work," I present a systematic and structured outline of the author's main works.

The purpose of this chapter is twofold: first, to offer a general overview of Hacking's work by means of a systematization of his bibliography, departing from the lesser-known axles of his work. Second, to look for an underlying texture in Hacking's work, departing from his interest in the conditions of possibility of the emergence of certain concepts and objects, and explore how Michel Foucault's thought works as a thread that runs across this web.

Based on a wide review and critical analysis of his bibliography, I aim to organize Hacking's work around four nodes. I show the internal structure of each node, I review the most representative texts, and I present the main notions with each one.

The chapter explores a series of relations between the nodes which, besides showing the reticular structure of Hacking's work, justify the order in which they are presented.

In Chap. 2, "Styles of Scientific Thinking & Doing: A Genealogy of Scientific Reason," I aim to work on the "style of scientific reasoning," later called "style of scientific thinking & doing" (Hacking 2009a).

I deal with this node in the first place because I visualize it as basal, both because it is a condition of possibility for the emergence of objects and concepts that appear in other nodes, and because one way or another all the others are related to it.

At the beginning of the chapter, I deal with the antecedents, differences, and convergences that Hacking's notion of style of scientific reasoning has with meta-concepts proposed by other philosophers and historians of science. I show how already in 1979, in an article where Hacking offers his interpretation of Foucault's thinking on the life sciences, work, and language, he discusses some of what will later be characteristic of his notion of style of scientific reasoning. I deal in particular with the notion of style of scientific thinking of Alistair Crombie, since Hacking considers him an immediate antecedent of his notion of style of scientific reasoning.

I continue by going deeper into the notion of style of scientific reasoning by means of an analysis of its particular features, as well as their relation to notions proposed by Foucault.

In Chap. 3, "Probability: Books That Smell of Other Books," I deal with the node of probability by means of an analysis of two of Hacking's most representative texts, and emphasizing the specific footprint of certain works of Michel Foucault.

Regarding *The Emergence of Probability*, I show that Hacking not only uses the Foucauldian archaeological method but he practically paraphrases its characteristics when explaining the methodology he uses in this book. I also show that while Foucault analyzed the historical conditions of possibility of knowledge for each one of the historical periods he deals with in *The Order of Things*, Hacking deals, mainly in the first six chapters of *The Emergence of Probability*, to show which are the historical conditions that make the emergence of probability possible.

The Taming of Chance is a book where Hacking aims to show that analytic philosophy does not need to be the antithesis of historical sensibility, and even though he uses the same archaeological methodology, he goes beyond it to reflect the influence of Foucault's genealogical works. This can be seen in what could be called an incursion into Foucauldian bio-politics. Hacking returns in this text to the idea that the organization of concepts and the difficulties that emerge from them sometimes have to do with its historical origins. Following Foucault's line called "history of the present," Hacking aims to understand how we think and why we seem to be compelled to think in a certain way.

In Chap. 4, "Making Up People: A Project of More Than Three Decades," I present the notions Hacking uses to work in the human sciences, inter-relating them to present the framework within which Hacking reflects about this realm of the sciences and I analyze the influence of Foucauldian thought in this area.

Hacking defines himself as a dynamic nominalist, insofar as he is interested in the interaction between classification and the classified individuals, and he vindicates Nietzsche and Foucault as antecedent of this nominalism. In particular, Hacking focuses on Foucauldian nominalism, being interested in the essential role of history in the constitution of its objects, people, and forms of behavior. Hence, his idea of historical ontology, which deals with the ways in which the possibilities of choice and of being emerge from history and from making up people, that is to say, the ways in which a new scientific classification can make a new kind of people emerge, conceived and experienced as a way of being a person. This interaction between classification and the classified individual results in what Hacking calls the looping effect of human kinds: the process of feedback suffered by human classes, due to the interaction between people and the forms in which they are classified. But making up people takes place in a space of possibilities that Hacking characterizes as an ecological niche, drawing inspiration from Foucault's notion of discursive formation.

Given that in the kinds of the human sciences the aforementioned loop effect presents itself, Hacking proposes the existence of different classes of kinds. In order to illustrate this, I recuperate Hacking's work process on this matter, departing from his original question about whether kinds of people are natural kinds, going through his distinction between natural kinds and human kinds based on the existence or not of the loop effect, following with the distinction between indifferent and interactive kinds, based on which Hacking posed his distinction between natural and human sciences. Hacking will eventually abandon this distinction and the use of the notion of interactive kind. This means, in my view, a great loss, although Hacking continues to vindicate the existence of an interaction between the kind and the classified.

Even though Hacking generally vindicates the influence of the archaeological stage of Foucauldian thought on his own thought, I believe that in many aspects there is a footprint of the genealogical stage of the French philosopher. In spite of this, Hacking does not deal with power, one of the characteristic concepts of Foucauldian genealogy, at least not in a systematic and explicit way.

In Chap. 5, "Classifications, Looping Effect and Power," I develop my idea that the aforementioned features are not an obstacle to our seeing the fundamental role of power in relation with the notions of classification, looping effect, and making up people. To this end, I start with a brief outline of the idea of power in Foucault, which I take as a starting point to show how the elements he takes as essential in a power relation can be clearly identified in the looping effect. I illustrate how power is imbricated in Hacking's proposal, resorting to several of the examples he himself has used to develop the notions mentioned in the title of this chapter.

Chapter 6, "Experimentation and Scientific Realism. A Return to Francis Bacon," is devoted to the fourth node: experimentation and scientific realism. Mainly based on *Representing and Intervening*, I present his criticism of representation, the theory, and realisms based on them. Hacking claims that for the most part, debates between realism and antirealism take place in terms of representation and theory, and as long as this happens, realism will not be capable of facing the challenges of antirealism. To correct this, he suggests shifting the discussion from realism to a different realm, experimentation, where—in his opinion—scientific realism is irresistible.

I continue by presenting his defense of entity realism. The central notion in Hacking's argument for scientific realism is not reference but manipulation. Electrons, for instance, only lose their hypothetical status when they are manipulated to investigate something else, when they stop being theoretical and become experimental. This engineering is, for Hacking, the best proof of scientific realism and the first argument of his defense: intervention.

The second argument claims that it would be a ridiculous coincidence if, once and again, two different physical processes produced visual configuration that were artefacts of these physical processes and not real structures of the entity. Hacking uses in this case the example of microscopes.

Hacking's incursion in the realist debates of the 80s decade was just a strategy to draw attention towards experimental scientific activity, to defend the vital and preeminent role of experimentation, and more precisely, the creation of phenomena in science.

In Chap. 7, "On Foucault's Shoulders," I start by proposing that the subject that awarded Hacking more visibility and consideration, dealt with in *Representing and Intervening*, is not the one that he has been most interested in nor the one to which he has devoted more time and publications. I show that when one analyzes his corpus as a whole, a very different image emerges from the one that appears if one focuses its study only on this area of his work.

This analysis, decentered from the axis around which the work of Hacking has traditionally been understood, is the basis of my proposal for this book. To defend it, I articulate the items dealt with in previous chapters and I close the circle I open in the first chapter, going back to the proposal of a reticular structure in Hacking's work and the place Michel Foucault occupies in this web, which I define as the texture, the always underlying thread that runs across in the loom where the cloth is woven. I claim that it is the notion of conditions of possibility for the emergence of concepts and objects, inherited and at the same time different from Foucault's, that impregnates Hacking's whole corpus.

Finally, in the "Epilogue," I claim that Hacking's proposal for the human sciences, mainly his notions on how to make up people, looping effect, and interactive classifications, not only can be complementary of Foucault's thought, but that this complementarity can give as a result a vision that goes beyond both projects considered separately and offer a better explanation of the objects of the human sciences and their behavior. This is the case insofar as Hacking's most concrete proposal would manage to complete Foucault's more abstract analyses, showing one of the mechanisms by means of which a relation is established between practices, institutions, and discourse on the one hand, and people and their behaviors in daily life on the other.

References

Álvarez Rodríguez, A. (2002). Entrevista con Ian Hacking. *Cuaderno de Materiales*, 17. http://www.filosofia.net/materiales/num/num17/Hacking.htm

D'Agostini, F. (2000). *Analíticos y continentales*. Madrid: Cátedra.

Foucault, M. (1966). *Les mots et les choses. Une archéologie des sciences humaines*. Paris: Gallimard.

Hacking, I. (1964). On the foundations of statistics. *The British Journal for the Philosophy of Science, 15*(57), 1–26.

Hacking, I. (1965). *Logic of statistical inference.* Cambridge: Cambridge University.

Hacking, I. (1975a). *The emergence of probability*. Cambridge: Cambridge University.
Hacking, I. (1975b). *Why does language matter to philosophy?* Cambridge: Cambridge University.
Hacking, I. (1981a). The archaeology of Michel Foucault. In I. Hacking (2002), *Historical ontology* (pp. 73–86). London: Harvard University.
Hacking, I. (1983a). *Representing and intervening*. Cambridge: Cambridge University.
Hacking, I. (1990a). *The taming of chance*. Cambridge: Cambridge University.
Hacking, I. (1998a). *Mad travelers. Reflections on the reality of transient mental illnesses*. Virginia: University of Virginia.
Hacking, I. (1999a). *The social construction of what?* Cambridge: Harvard University.
Hacking, I. (2002a). *Historical ontology*. London: Harvard University.
Hacking, I. (2009a). *Scientific reason*. Taiwan: National Taiwan University.
Kuhn, T. (1962). *The structure of scientific revolutions*. Chicago: University of Chicago.
Vagelli, M. (2014). Ian Hacking. The philosopher of the present. *Iride, 27*(72), 239–269.
Vattimo, G. (2000). Prefacio. En F. D'Agostini (2000), *Analíticos y continentales* (pp. 13–17). Madrid: Cátedra.

Hacking, I. (1975). *The emergence of probability*. Cambridge: Cambridge University Press.
Hacking, I. (1975b). *Why does language matter to philosophy?* Cambridge: Cambridge University Press.
Hacking, I. (19[illegible]). The archaeology of Michel Foucault. In I. Hacking (2002), *Historical ontology* (pp. 73–86). London: Harvard University Press.
Hacking, I. (1983). *Representing and intervening*. Cambridge: Cambridge University Press.
Hacking, I. (1990). *The taming of chance*. Cambridge: Cambridge University Press.
Hacking, I. (1998). *Mad travelers: Reflections on the reality of transient mental illnesses*. [illegible]: University of Virginia.
Hacking, I. (1999). *The social construction of what?* Cambridge: Harvard University Press.
Hacking, I. (2002). *Historical ontology*. [illegible]: Harvard University Press.
Hacking, I. (2009). *Scientific reason*. Taipei: National Taiwan University Press.
Kuhn, T. (1962). *The structure of scientific revolutions*. Chicago: University of Chicago Press.
Vagelli, M. (2014). Ian Hacking: The philosopher of the present. *Iride*, 27([illegible]), 239–269.
Vesterinen, T. [illegible] [illegible] [illegible]
[illegible]

Contents

Chapter 1
"Taking a Look" at Ian Hacking's Work

> *A philosopher I will not name complained once: 'philosophers never take a look at what they discuss'. Well, not completely, Ian Hacking does take a look*
> *Ian Hacking*
>
> (Álvarez Rodríguez 2002:8)

Abstract In this chapter, *Taking a look at Ian Hacking's* work, I present a novel perspective on Ian Hacking's oeuvre, developing a systematic and structured outline of the author's main works. The purpose of this chapter is twofold: to offer a general overview of Hacking's work by means of a systematization of his bibliography, departing from the lesser-known axles of his work. Secondly, to look for an underlying texture in Hacking's work, departing from his interest in the historical and situated conditions of possibility of the emergence of concepts and objects scientific, and explore how Michel Foucault's thought works as a thread that runs across this web. Based on a wide review and critical analysis of his bibliography, I aim to organize Hacking's work around four nodes: styles of scientific thinking & doing; probability; making up people and experimentation and scientific realism. I show the internal structure of each node, I review the most representative texts and I present the main notions with each one. The chapter explores a series of relations between the nodes which, besides showing the reticular structure of Hacking's work, justify the order in which they are presented.

Keywords Ian Hacking's oeuvre · Nodes · Styles of scientific thinking & doing · Probability · Making up people · Experimentation and scientific realism · Michel Foucault

M. L. Martínez Rodríguez, *Texture in the Work of Ian Hacking*, Synthese Library 435, https://doi.org/10.1007/978-3-030-64785-8_1

1.1 The Nodes of the Network

In an interview with Asunción Álvarez Rodríguez, Ian Hacking answered the question about whether he saw "some thread or threads that ran across his work," saying that he had never felt a great need to unify his work, but that of course he saw strong connections between the different parts of his work (2002: 8). He later added that these parts, rather than constituting a single underlying theme, form a network of connections and he gave as an example some fragments of that network: the subject of probability, that of making up people, and Michel Foucault.

According to my reading of Hacking, however, his work is not divided in watertight parts, but rather, in that reticular structure, the apparent sections of his thought actually function as nodes, among which I identify the following: (1) style of scientific reasoning or of scientific thinking & doing, (2) probability, (3) making up people, and (4) experimentation and scientific realism.

Why these changes? In the first place, whereas Hacking speaks of parts of the network that makes up his work, I prefer to call them "nodes", in the sense that they are abstract spaces of confluence of part of the connections of other abstract spaces that share some of their features and that are themselves nodes. They interrelate in such a way that they make up a network, which can be defined as a set of interconnected nodes. These features of the nodes and their relation with the network are what justifies my choice, insofar as I believe they better reflect the structure of the Canadian philosopher's thought.

Secondly, I add two nodes that Hacking does not mention. The style of scientific reasoning and experimentation and scientific realism. Surely he did not mention them because he was not making an exhaustive enumeration of his work, but rather naming some of its parts as an example. With regards to the last node I must remark from the beginning that, unlike its traditional interpretation, the topic of scientific realism is not for Hacking what really matters. Scientific realism was for him only "*a topic in vogue at that time,* [which] *was convenient to attach to a book that nobody seemed to want*". (Álvarez Rodríguez 2002:8)

Thirdly, I must explain why I prioritize the node of style of scientific reasoning. Hacking has claimed (2009b) that his long-term project are the styles. In this sense, it is possible to see Hacking's work as structured with this topic as a core from which a series of related and inter-related themes emerge, beyond the fact that some of these notions have acquired a level of development such that it is possible to identify them as other important nodes in the network. Let us remember that the probabilistic style is an exemplar of the style of reasoning; that the idea of making up people is the result of his work on the statistical style, showing how classifications—dependent on the style—give way to new objects and kinds, particular of each style, and that scientific realism-antirealism debates take place within each style, insofar as they are producers of their own objects.

Fourthly, and most importantly, Michel Foucault cannot be considered simply another node, in the same sense as the style of scientific reasoning, probability, making up people and experimentation and scientific realism.

My proposal is that the influence of the French philosopher is clearly present in every portion of the network and it constitutes the matter that makes up the texture of most of Hacking's work.

As I have already remarked and will become clearly in later sections, the portions of the network inter-connected in such a way that they not only show Hacking's particular way of working, "taking a look" at the rich complexity of the world—without setting out, in general, to develop a conceptual analysis of this richness in the typical style of analytic philosophies—but they also suggest the existence of some thread that runs across his work, as well as certain interests that are kept throughout it.

1.1.1 *Style of Scientific Reasoning or Thinking & Doing*

With regards to the "style of scientific reasoning or thinking & doing", since undertaking this project in 1981 with his well-known article "Language, Truth and Reason" (1982), Hacking has published two books, *Scientific Reason* (2009a) and *Why is There Philosophy of Mathematics at All?* (2014), as well as several articles, including "'Style' for Historians and Philosophers" (1992b), "Statistical Language, Statistical Truth and Statistical Reason: the Self-authentification of a Style of Scientific Reasoning" (1992c), "The Self-vindication of the Laboratory Sciences" (1992a), "Styles of Scientific Reasoning: a New Analytical Tool for Historians and Philosophers of the Sciences" (1994).

However, and even though this notion does not appear explicitly, it could be said that already in *The Emergence of Probability* (1975a)—strongly influenced by Michel Foucault—there are traces of Hacking's thought on this subject. This book will be discussed as part of the probability node.

In "Language, Truth and Reason", Hacking remarks that he elaborates the notion of *style of scientific reasoning* on the basis of that coined by Alistair Crombie: *style of scientific thinking*. The essential difference between both has to do with the fact that Hacking wants his concept to refer not just to thinking but also the hand that manipulates; in other words, not only to thinking but also to doing, not just to representing but also to intervening, not just to the mental and private but also the public, and for this reason he prefers to use the term reasoning rather than thinking.[1] In his following papers, Hacking will continue to characterize this notion of style of scientific reasoning, which he defines as "[…] *a new analytical tool that can be used by historians and philosophers for different purposes*" (Hacking 1992b: 1). In this

[1] Whereas Hacking (1992b: 4–5) states that his choice of the term reasoning is related to this and with its Kantian echoes, my perception is that it does not completely fulfil Hacking's objective: to include not just the "thinking" but also the "doing" of scientific practice. Neither does Hacking seem to be satisfied by this choice, since in 2010 he proposes to substitute his expression "style of scientific reasoning" by "style of scientific thinking & doing", which in my view represents more accurately Hacking's objectives and interests.

context, he points out that the style is a durable, impersonal social unit, a way of seeing and acting, and it does not determine a specific content or science but is common to several. The style introduces a number of innovations: objects, elements of proof, laws, possibilities, kinds of classification and explanation characteristic of each one of them. It also establishes what statements can have truth value, and it has its own stabilization techniques.

As well as the examples of the styles from the history of the concepts of probability and indeterminism in *The Emergence of Probability* and *The Taming of Chance* (1990a) Hacking devotes parts of other works to this exemplification. In the second part of "Statistical Language, Statistical Truth and Statistical Reason: the Self-authentification of a Style of Scientific Reasoning", he analyzes the development of the statistic style, pointing out which event mark its different stages, when and how new sentences, laws, objects, explanations, criteria, etc. appear.

"The Self-vindication of the Laboratory Sciences" deals with the other main example proposed by Hacking to illustrate his idea of scientific reasoning: the laboratory style. In this work he remarks on how laboratory science justifies itself by alluding to internal (ideas, things, symbols) and external elements, following the idea that a mature laboratory science develops a mutually-adjusting body of kinds of theories, apparatus and analyses (1992a:30).

In this text, the idea of incommensurability also appears, not in terms of meaning but in terms of laboratory practice, since commensurability is impossible because the instruments that provide results for one theory do not fit another.

In 2009 Hacking published *Scientific Reason*, the product of a series of conferences he had given in Taiwan two years earlier. In this text the author refers to the historical roots of scientific reason, in order to emphasize that the human being has learned not only about the world and how to change it, but also how to research, by developing social organizations within which to promote certain innate skills.[2] Cognition and culture are two dimensions that offer a space to understand scientific reason. Hacking is interested in a long-term, philosophical and anthropological vision of scientific reason. To this end, he returns to the origins of his idea of style, dating back to Crombie, but abandoning his expression of style of scientific reasoning—coined to distinguish his own notion from that of style of scientific thinking of the Australian historian—and he substitutes it by style of scientific thinking & doing. He remarks that his main innovation (inspired by Foucault) with regards to Crombie is the idea that styles, which are long-lasting, are interrupted by what he calls "crystallizations". He deals with the subject of scientific truth taking as a point of departure the notion of truthfulness, conceiving it as dynamic and historical rather than static and ahistorical, based on Bernard Williams'[3] genealogic investigation. He also discusses the realism-antirealism debate, defending that these debates

[2] Innate abilities such as, for example, "*Logic, in the sense of Peirce's triad* [deduction, induction, abduction] *is a human universal* […]" (Hacking 2009a: 99).
[3] See Williams (2006).

are a product of the introduction of objects in the styles of scientific thinking & doing. Finally, he devotes a good portion of his book to discuss the laboratory style.

Finally, in his last work, *Why is There Philosophy of Mathematics at All?*, whose central subject is the question of the title, Hacking deals with questions related to mathematics, one of the six styles of scientific thinking & doing that he considers to be an integral part of modern scientific activity. However, this book is not *stricto sensu* illustrative of the aforementioned style, but rather a reflection on why there is a perennial philosophical interest in mathematical questions. The birth of mathematics can be understood as the discovery of a capacity of the human mind or thinking, hence its great importance to philosophy. And the reasons for this persistent interest can be found in two threads, one which comes from antiquity, whose emblem is Plato, and another that is revealed in the Enlightenment, whose icon is Kant. Throughout Antiquity, this fascination emerges from the idea that mathematics explores something "out there". During the Enlightenment, although the point of departure is proof, the focus is on the certainty of that which is proven, the experience of absolute necessity, of apodictic certainty. In Antiquity, the mathematical experience, proof, is an essential part of the discipline since Plato, a certain guide towards truth; this is a contingent historical fact, it could not have been developed or it could have developed otherwise. In the Enlightenment, the core is the physical experience, application or use. Applications, in their six kinds,[4] foster or generate their own philosophical dilemmas, originated in the Enlightenment and different from those derived from the Antiquity thread, thus contributing to the perennial and central character of the philosophy of mathematics in the western philosophical tradition. In this work, Hacking conjugates old vindications such as the *doing* in science—mathematics are conceived by Hacking as activities grounded in the body, both on the hands and on the brain, carried out by human communities in specific times and places, and which, when applied, can be used in the material world; Ludwig Wittgenstein's influence[5] -who remarkably emphasized the different

---

[4] *App 0: Maths applied to maths:* for instance, when Descartes applies algebra and arithmetic to geometry to create analytic geometry; *App 1: The Pythagorean Dream*: where it is considered that the essence of the universe is a mathematical structure; *App2: Mathematical physics:* the elegant mathematical structures provide surprisingly precise models of the processes found in nature; *App 3: Mission-oriented applied maths:* it covers what is frequently considered applied mathematics; its prototype is the Society of Industrial and Applied Mathematics (SIAM); *App 4: Common or garden,* innumerable common uses of maths by merchants, accountants, financiers, farmers, carpenters, etc. *App 5: Unintended social uses:* unforeseen uses of mathematics, for example, the elitist use of mathematics by Plato; *App 6: Bizarre applications:* according to Hacking, these are uses of mathematics not only unforeseen by the mathematicians who do the work, but which would not have occurred to us but for the legacy of Wittgenstein.

Hacking considers that these distinctions between different types of applications are not clear-cut and the series should not be taken as implying a linear, progressive accumulation. What he wishes to emphasize is how the familiar expression 'applied mathematics' can mask such variety of uses (Hacking 2011a: 156–158; 2011b: 11–14).

[5] Hacking makes it clear, however, that he does not intend to be an interpreter of Wittgenstein, and even that he does not sympathize with everything written by the Austrian philosopher. He learns from reading Wittgenstein and incorporate that which he reads in his own idiosyncratic way, in his

activities that understood as mathematics-; and ideas that emerged in more recent texts about styles of scientific thinking & doing, such as his proposal that the human being has certain cognitive capacities, certain inherited abilities that have evolved and which, exploited in a certain time and place by a group of people, became part of the system of the world, in our standards of 'good reasons' (Hacking 2014: 142).

1.1.2 Probability

On the topic of probability, four books stand out: *Logic of Statistical Inference* (1965), *The Emergence of Probability*, *The Taming of Chance* and *An Introduction to Probability and Inductive Logic* (2001), as well as a series of articles written mainly between the decades of 1960 and 1990, among which I would like to underscore "How Should We Do the History of Statistics?" (1981) and the aforementioned "Statistical Language, Statistical Truth, and Statistical Reason: the Self-Authentification of a Style of Scientific Reasoning".

This meta-philosophical node, which partly precedes chronologically the other ones mentioned here, can be seen fundamentally as a detailed historical background, which resulted in a substantial part of Hacking's later investigations.

Logic of Statistical Inference, considered by its author as an important contribution to the philosophy of science, is a philosophical analysis applied to probabilistic reasoning. It was the first attempt by a philosopher to analyze the ways in which statisticians extract inferences from data. Theoretically, it is framed in the debate between frequentists and neo-Bayesians. Hacking attempts here to construct an axiomatic theory of statistical inference, grounded on the frequentist point of view and able to respond rigorously to the question posed by the neo-Bayesians.

To leave the enclosed world of philosophy, Hacking resorts then to the historical detail,[6] focusing on and reconstituting from its origins the probabilistic thought between 1654 and 1737, resulting in *The Emergence of Probability*. Here, he shows how during the seventeenth century, and as a result of a radical change that took place very rapidly from the previous Renaissance conceptions, a probabilistic reasoning was developed, in opposition to a deterministic view of reality. "[…] *around*

---

own philosophical thought. In this case, Hacking takes some of the lines of thought which appear in *Remarks on the Foundations of Mathematics*, because he considers that it suggests an alternative to the traditional philosophical conceptualization of mathematics and what it deals with, which has the benefit of not raising certain philosophical perplexities which seem impossible to overcome in the traditional conceptualization (Hacking 2011a: 155; 2014: 59).

[6] Hacking reminds us in an interview (Vagelli 2014:247) that his first approach to probability was along the lines of the linguistic analysis of the word "probability". He still believes that the text he wrote about it was probably correct, but he did not consider it relevant. For this reason, he changed the key of his linguistic analysis for a historical analysis. However, it can be said that something of this pure and profound linguistic focus remained in the book of 1975. The result of this rigorous linguistic analysis were the two articles he wrote on possibility in the *Philosophical Review* at the end of the 60s and beginnings of 1970s (Hacking 1967 and 1975b).

1660 a lot of people independently hit on the basic probability ideas. It took some time to draw these events together but they all happened concurrently" (Hacking 1975a: 11–12).

His central idea is that the specificity of the notion of probability is the duality and the tension that recur among the two aspects of this "double head of Janus". On the one hand, in epistemic terms, it aims to evaluate reasonable degrees of belief (warranted by factual evidence).[7] On the other hand, in statistical terms, it is connected with the tendency exhibited by some dispositives to produce stable, long-term relative frequencies. Hacking follows the tracks of this idea of duality through the stages enumerated between 1654 and 1678, defending the idea that the notion of probability that emerged in the seventeenth century preserves this initial duality to our days.

Even though Foucault is explicitly mentioned on only two occasions, the influence of *The Order of Things* is evident in this book, not only in passages like the following:

> There is an anti-positivist model which, for all its obscurity, may at this point have some appeal. We should perhaps imagine that concepts are less subject to our decisions than a positivist would think, and that they play out their lives in, as it were, a space of their own. If a concept is introduced by some striking mutation, as is the case with probability, there may be some specific preconditions for the event that determine the possible future courses of development for the concept. All those who subsequently employ the concept use it within this matrix of possibilities. (Hacking 1975a: 15).

This presence is also evident when Hacking describes the methodology he uses in the research that gives place to the text, which I will analyze in a later chapter.

The Taming of Chance, for its part, deals with the eighteenth and nineteenth centuries, showing that the recollection of numbers and the growth of statistical analysis led twentieth century philosophers to abandon a mechanistic view of the world and to adopt another one based on chance. This book, the result of Hacking's participation in a seminar on the probabilistic revolution (1982–1983), develops ideas already present in the article of 1981. His central idea is that the "taming of chance" results from the application of the law of big numbers, which made it possible, from the years 1820–1830, to think about the world in non-deterministic terms on a microsocial level, within the framework of a probabilistic statistical model that develops the consequence of the idea of stable relative frequency presented in the two previous books. It is after the treatment of social material such as bureaucracy, crime, suicide, illnesses, the thoracic measurements of soldier's chests, etc., that the probabilistic world was developed. Like in *The Emergence of Probability*

[7] The problems of translation of the word *evidence* from English into other languages such as French, led Hacking to make some clarifications on his use of the term. In English, *evidence* is a false friend of his French homonym *evidence*. In French, the evidence is mainly intellectual. The word comes from *videre,* to see, but in French it means to see with the spirit, Descartes' intuition. In English, *evidence* means the facts, the data, which indicate other facts, and sometimes the positive proof of these facts. In the Preface to the French edition of *L'émergence de la probabilité* (2002b), Hacking remarks that in chapter 4 of *The Emergence of Probability* (1975a), he uses the term in this English sense of factual evidence.

(1975a:27–28), Hacking points out here that only once this was done in the world of the human could physics enter the field of probabilities, when the social thought had already blazed a trail. The subject of the taming of change is accompanied in *The Taming of Chance* by others such as: the erosion of determinism, the autonomization of statistical laws and the so-called "avalanche of numbers".[8] By means of this incursion into political-administrative history Hacking shows the links between, on the one hand, the evolution of the use of statistics and the emergence of an autonomous statistical style of reasoning, and on the other, the transformations of the State, where the organization of bureaus of official statistics makes the avalanche of numbers possible, and for this reason, the incorporation of long-term frequencies in the normal daily practices of institutions and social actors. By underscoring the role of statistical bureaus, Hacking faces the problems of classification and nomenclature that he will develop in later works, about the notion of kinds in general, and mainly of kinds in the human sciences.[9]

In the last chapter, and apropos C.S. Peirce, the issue of induction appears to close—at least provisionally—the gap opened by his questions about its grounding in his 1965 book.

The three aforementioned books are devoted to understanding the relation—or absence thereof—between inference in the Bayesian manner and the kinds of inference called objective, which are based on the notion of frequency. All three conclude, besides, with an examination of the problem of induction attributed to Hume.

The book published in 1990 was followed by the paper of 1992, centered in looking for a connection between certain questions that Hacking had been analyzing in his works on probability, as well as other more general ones, postulated by the philosophers of science, about entity realism. There he develops the notion of style of reasoning, applying it to statistical reasoning. The context of this question is provided by the philosophical task of connecting the three following orientations of investigation: (1) the social studies of knowledge, of which David Bloor and Barry Barnes are pioneers in the Anglophone tradition; (2) metaphysical reflections, particularly the debates resulting from the revision of some positions such as Hilary Putnam's, who begins with a defense of metaphysical realism and then rejects to embrace internal realism, and (3) a Braudelian conception of science, with its belief in the long term and the persistent and cumulative growth of knowledge. To bridge

[8] In the period of peace following Napoleon, European states created bureaus to gather and publish statistics on all aspects of life and administration. These bureaus made possible the avalanche of printed numbers, from 1820 to 1840. *Recherches statistiques de la ville de Paris et le département de la Seine* was the beginning of what was called the avalanche of printed numbers.

[9] Hacking claims to choose this expression –with clear connotations in French but not systematically used in English—because it includes many social sciences, psychology, psychiatry, and a good part of clinical medicine. However, years later he claims that the so-called natural sciences include medicine, cognitive psychology and positive sociology, among other disciplines (2010, April 21: 3). Hacking does not clarify the reason of this change of mind, but I think it might be related with the latest results of his investigations about the distinction between the kinds of natural and human sciences, and his idea that, finally, there would not be a kind of human kinds to oppose to the kind of natural kinds.

this gap Hacking proposes the style of reasoning. This emerges from social and scientific practices, but then becomes autonomous with respect to the conditions in which it emerged, acquiring maturity and becoming stabilized.

In *An Introduction to Probability and Inductive Logic*, Hacking resumes the main subjects of his previous works. He deals with probability and induction. He includes numerous historical episodes about how both ideas developed, and he offers, after all the path travelled, his opinion about the structure of these logical ideas. Hacking himself has pointed out the importance of the final chapters of the book as a contribution to the philosophical problem of induction. In this sense, there is a close connection between this book and the two previous ones: *The Emergence of Probability*, where a central topic is how the problem of induction became possible, and *The Taming of Chance*, which, as we have already mentioned, culminates with C.S. Peirce and his understanding of inductive inference. Moreover, in the Preface to the latest edition of *The Emergence of Probability* (2002b), Hacking remarks that the only quote by Michel Foucault that appears there ends by saying: "*Hume has become possible*". This confirms, according to the author himself, that the underlying thread of his research on probabilities is the way in which Hume and the problem of induction become possible.

The book of 2001 also deals –albeit indirectly—with an essential topic for the statistician: categorization, through his questioning on kinds.

Finally, the concern about the way in which we live in a "random universe", i.e. the way in which many of our worries and decisions have fallen into the realm of probability, is another constant that runs across all this part of his work. In the first chapter of *The Taming of Chance* he remarks:

> Probability and statistics crowd in upon us. The statistics of our pleasures and our vices are relentlessly tabulated. Sports, sex, drink, drugs, travel, sleep, friends –nothing escapes (Hacking 1990a: 4).

And in *An Introduction to Probability and Inductive Logic*, he claims: "*Nowadays you can't escape hearing about probabilities, statistics, and risk. Everything –jobs, sex, war, health, sport, grades, the environment, politics, astronomy, genetics– is wrapped up in probabilities*" (Hacking 2001: xi).

1.1.3 Making Up People

This topic includes texts such as *Rewriting the Soul. Multiple Personality and the Sciences of Memory* (1995a), *Mad Travelers. Reflections of the Reality of Transient Mental Illnesses* (1998), and *The Social Construction of What?* (1999a), as well as numerous articles among which I wish to mention especially "Making Up People" (1986), "The Looping Effects of Human Kinds" (1995b), and "Historical Ontology" (1999b).

Even though the first important texts—at least in terms of extension—appear only in the mid-1990s, Hacking has published on the subject almost without interruption at least since 1982. Considering not just the texts where this is the central subject but all those related with it and on the issue of the social sciences, it turns out that the bibliography produced in this field widely exceeds that for which he has become better known as a philosopher: that related to his position on experimentation in the natural sciences and on scientific realism.

In 1986 Hacking published the paper "Making Up People", where he attempts to answer a series of questions related to ways of classifying people and the consequences that these classifications have on them. Departing from a fragment from Arnold Davidson that says that perversion was not an illness that lurked in nature waiting for psychiatrists to discover it but an illness created by a new (functional) understanding of the illness, a conceptual change, a change in reasoning, which made it possible to interpret several types of activity in medical-psychiatric terms, Hacking analyzes if there could be a general theory about the making up people or if each example is so peculiar that it demands its own history. He also wants to know how this idea of making up people affects our idea of what it means to be an individual. In the development of his analysis there emerge some of the notions that will constitute the present node: making up people and dynamic nominalism, and he elaborates on the Foucauldian notions of anatomo- and bio-politics.

Nine years later, "The Looping Effects of Human Kinds" deals with how a classification, if it is known by those people it is applied to, can change the kind of persons they are. This can in its turn lead to a change in the way of classifying them, since the old form is no longer adapted to the new features of the members of the kind. Thus emerges the notion of the looping effect.

The two following books can be considered as two chapters in Hacking's commitment with the history of psychiatry. Originally, *Mad Travelers* was meant to be a chapter of *Rewriting the Soul,* but it turned out to be too long, and Hacking was fascinated by the history of a very strange man, Albert Dadas, the original *fuguer*. Both books can be read as philosophy books, in relation to the possibility of investigating the historical mechanisms according to which new kinds of people emerge, dealing with classification and generalization, with doing and thinking, with a particular sense of nominalism, with the defects of an exclusively linguistic approach to these problems. It is also possible to find in them a kind of ontological debate, raised by the question of the reality of mental illnesses, whose existence is due more to the social environment than to neurological defects.

Rewriting the Soul analyzes the emergence of the sciences of memory, the new meaning of the concept of trauma and the emergence of multiple personality disorder (MPD), as well as child abuse as concepts and as objects of knowledge. There Hacking outlines the history of MPD diagnosis from its beginnings, its disappearance in favor of schizophrenia at the beginning of the twentieth century, and its popularity, particularly when used in the defense of criminal cases. He is interested in elucidating the conceptual background that made possible the debates on memory and abuse, as well as the appearance of the disorder in nineteenth century France, when memory supplanted the soul in the scientific study of personality. In this context, the concept of trauma was crucial to make of memory a scientific problem between 1874 and 1886.

This book resorts once more to the archaeological strategy already used in his two previous books, as the author himself remarks in its first pages (Hacking 1995a:4). But as well as the influence of *The Order of Things*, this work also shows a distinctive mark of *L'archéologie du savoir* (1969), translated into English as *The Archaeology of Knowledge and the Discourse of Language*. It constitutes a memory-politics of the soul in the mode of the anatomo-politics of the body or the bio-politics of the population in Foucauldian terms. He considers that sometimes there are mutations in the systems of thought, and these redistributions of ideas establish what later seems inevitable, unquestionable and necessary. This is what has happened with multiple personality disorder, whose latest events have been possible thanks to the development of the field of knowledge about memory.

In this text, Hacking approaches multiple personality led by an interest that had already appeared in *The Taming of Chance*: the making up of people, individuals that are constructed by interacting with the classification that the experts have submitted them to. The concern about how kinds of people emerge, how the systems of knowledge about kinds of people interact with the persons known under these kinds. Both if this is the creation of an absolute moral—as in child abuse—or the use of statistics to define the given, Hacking has carried out important historical-philosophical investigations on how what we construct determines who we are.

The interest in making up people and on how the scientific notions of kinds of people affect the members of the class, continues, as we have mentioned, in his later book: *Mad Travelers*, since both were motivated by the desire to understand how and when new kinds of people appear. The case that this other text examines is that of Albert Dadas, an employee of the gas company in Bordeaux who one day abandons his house, his job and his daily life to start a long journey. He travels compulsively until he is arrested for vagrancy, imprisoned and then returned home, without any idea of where he had been and remembering confusedly and only in a hypnotic state what had happened.

After 1887 an epidemic of "compulsive travelers" or *fugueurs* is verified. Experts call it the syndrome of ambulatory automatism. Even though it had been known forever, only in this moment did the fugue become an illness in the psychiatry manual. Is it then a real mental disorder? Or an artefact of psychiatry? It is not Hacking's interest in this text to discuss if the illness is real or constructed, but rather to provide an outline to understand the possibility that the so-called transient mental illnesses.[10] Albert Dadas fugue appears precisely as a psychiatric syndrome because in those times there exists only the middle-class enthusiasm for tourism and/or

[10] Hacking defines transient mental illnesses as those that appear in certain moments and places and then disappear. As examples he mentions, among others, hysteria in France at the end of the nineteenth century, multiple personality disorder, more recently, in North America, and anorexia in the present (Hacking 1999a: 100). Whereas Hacking seems to claim that hysteria has disappeared, and we assume that it is in this sense that he considers it a transient mental illness, as it will be seen in a later note, agreement on its disappearance is far from unanimous. Moreover, Hacking is also aware that there are illnesses that are transitory for biological reasons. He is not interested in them, but only in those that do not have, in principle, biochemical, neurological or bacteriological explanations of why they appear.

generalized vagrancy. Without financial-social conditions to travel as a tourist and since he was not a criminal, there was no other classification available for Albert than that of illness. But this requires not just a medical taxonomy[11] but also that the illness could be observed and that it provides some sort of liberation that cannot be found elsewhere in culture. These are some of the vectors that make up the ecological niche that allows for the emergence of illness, the classification and subsequent new kinds of people.

On this occasion, Hacking mentions more than once his debt to Foucauldian thought, since the metaphor of the ecological niche is inspired in that of linguistic discourse of the French philosopher. This text, like others can be read as a slippage from the stricter analytic tradition towards Hacking's interest in the history of concepts and ideas, a product of his reading of Foucault. According to the Canadian philosopher, history, the archaeology of knowledge, can contribute to explaining certain phenomena, beyond logical analysis. They can help to do that which analytic philosophers, such as Hacking, are trained to do: to make distinctions and clarify ideas (1998:11).

They are trained to do this because they think it helps to remove conceptual confusion. But the most important conceptual difficulties are underlying ones. We have an unlimited reservoir of ignorance, but also conceptual confusions that new knowledge rarely helps to mitigate. There are several reasons for this, but in this particular case, the most interesting one according to Hacking, has to do with how scientific knowledge about ourselves change the way in which we think about ourselves, the possibilities presented to us, the kind of person we choose to be. Knowledge interacts with us and with an extensive body of practices and with daily life. This generates socially permissible combinations of symptoms and illnesses, as the text illustrates. A text that, in its structure, with a series of documents annexed to the end with the objective of providing documentary support to the data mentioned throughout, resembles Foucault's (1972) *Histoire de la folie à l'âge classique* (*History of Madness)*.

The subject of these two books will be taken up in *The Social Construction of What?*, a text whose leitmotif is to analyze the abuse—rather than the use—of the expression "social construction". Hacking returns here to some of the cases discussed in previous works, such as child abuse, already dealt with, besides *Rewriting the Soul*, in a series of articles from the late 1980s and early 1990s. However, there appear here not only subjects of the past but others that Hacking will develop from then on, such as autism and racism. He analyzes the way in which these kinds of people have been constructed, how the concepts and the objects have changed throughout history, apropos his explanation on what he calls interactive kinds, that belong to the human sciences and are opposed to the indifferent kinds of the natural sciences. This is framed within a discussion of the human sciences and their peculiar relation with their objects; the discussion of interactive kinds and the looping

[11] "Taxonomy" is understood not only as a systematic classification, but also a hierarchical one. In this work we have preserved the use Hacking in each particular case of the notion of classification or taxonomy, respectively.

effect through which persons can react consciously to the descriptions made by these sciences, acting on such descriptions and thereby forcing their revision.

This set of notions that has been forming around the concept of making up people—looping effect, interactive kind, ecological niche—is completed by another notion, that of historical ontology, which appeared in a homonymous article of 1999, and which names a collection of revised essays, written by Hacking between 1973 and 1999. Two tightly related subjects dominate this volume: some original senses in which the philosopher can make use of history and Hacking's early use of Foucault's "archaeology". The first is centered in the historical emergence of concepts and objects through new uses of words and statements in specific ways, and new models or styles of reasoning.

There are objects that have a historical ontology, they *"[...] do not exist in any recognizable form until they are objects of scientific study"* (Hacking 2002a:11). Hacking uses this idea to argue once more that, in the human sciences, the creation of classes frequently changes reality in a sense that does not apply to the natural sciences. The looping effect of human kinds is responsible for the fact that these objects, insofar as they are historically constituted, are in constant change. The consequence of this is the impossibility of working with a traditional kind of nominalism and the need, according to Hacking, for a dynamic nominalism –another notion that appears directly or indirectly in all these texts, and which has been developed by Hacking since the 1980s.

The subject of nominalism, as well as that of kinds, have interested Hacking with his project on the making up of people, but they acquired an independent development which, without being considered a node such as those discussed here, resulted in a series of publications.

With regards to nominalism, this subject was developed mainly to ground his proposal of a dynamic nominalism in the human sciences—in this case his most representative articles are "Five Parables" (1984) and "Historical Ontology"—even though the problem is present not only in his works devoted to the aforementioned sciences. He paid special attention to the proposal of the nominalist Nelson Goodman, apropos which he wrote a book that is only known in its French version: *Le plus pur nominalisme. L'enigme de Goodman: "vleu" et usages de "vleu"* (1993b). In this work Hacking analyzes what is philosophically at stake in Goodman's book *Fact, Fiction and Forecast* (1983). He frames the author's proposal in the nominalist tradition of Ockham, Hobbes, Locke, Hume, Mill and Russell, and he claims that ignorance of such tradition is what has led to the difficult understanding of the enigma of induction.

With regards to kinds, his interest dates back to 1980—as I said, surely apropos his work on the making up of people, but also his analytic training—and resulted in the publication of a series of articles about natural kinds and human kinds, the most outstanding of which are "Natural Kinds" (1990b), where he defends the idea that there are no classes in nature but that they are a product of a human construction; "A Tradition of Natural Kinds" (1991), where he reviews the tradition of natural kinds; "Working in a New World: the Taxonomic Solution" (1993a) in which apropos Kuhn's idea that scientists, after a change of paradigm, work in different worlds,

Hacking deals with nominalism, natural kinds, and presents a taxonomy —or rather, an anti-taxonomy—of his own. In *The Social Construction of What?* and in the article "How 'Natural' are 'kinds' of Sexual Orientation?" (2002c), Hacking works on the distinction between the kinds of the natural sciences and those of the human sciences, and in "Natural Kinds: Rosy Dawn, Scholastic Twilight" (2007) he goes back to his idea that there are so many competing views about what natural kinds are that a stipulative definition that took some classes and defined them as such would be purposeless. Even though it can be considered that some classifications are more natural than others, and despite the honorable tradition of class and natural kind, which dates back to 1840, there is no class, precise or fuzzy, of classifications that could be called, in any useful way, the kind of the natural kinds.[12]

1.1.4 Experimentation and Scientific Realism

The subject for which Ian Hacking's work has been known in the field of the Anglo-Saxon philosophy of science, his treatment of experimentation and scientific realism in the realm of the natural science is not, as I have remarked, the most extended in time nor the one that occupies more room in his bibliography. His works on this subject are concentrated mainly between the decades of 1980 and 1990. In a more recent interview, Hacking has talked about his interest in experimental physics, following his 1983 project, which resulted in the publication of his most important work in this area: *Representing and Intervening* (1983), a manifesto in favor of the study of experimental science and the role of experimentation in science. According to the author, the aim of this work was to invert the traditional hierarchy of theory over experiment, and to show that experimentation has a life of its own, independently of theory. The book is divided into two parts, each one devoted to the topics of representing and intervening, respectively.

Already in its first pages, Hacking starts his defense of realism of entities, contrasting it with a theoretical realism, based on representation and incapable of winning the battle against anti-realism. The first part of the text is wholly devoted to criticizing the exaggerated emphasis that twentieth century philosophy of science placed on theory and representation to the detriment of experimentation. The second part, intervening, offers arguments in favor of a realism of entities: intervention and coincidence. Through his two most famous examples—the electron and microscopes, Hacking illustrates the importance of doing in scientific practice, which allows for the creation of phenomena and favors to a great extent a strong scientific realism.[13]

These themes are central in few other articles: "Experimentation and Scientific Realism", written in 1982 and coinciding to a large extent with chapter 16 of the

[12] I do not refer with this expression to Russell's paradox.

[13] See Martínez (2009).

1983 book; "The Participant Irrealist at Large in the Laboratory" (1988), where he criticizes constructivism and Bruno Latour's proposal, and "Extragalactic Reality: the Case of Gravitational Lensing" (1989), which deals with realism and antirealism in astrophysics; but they are implicated in a great part of his work in some sense or other. Thus, when discussing transient mental illnesses, Hacking has claimed that there are two familiar questions in all the sciences: intervention and causation (1995a:12).

1.2 Constructing the Network

A systematic analysis of Hacking's work shows that, beyond the apparent dispersion that can be perceived on a first encounter, it responds as a whole to certain well-defined interests of the philosopher, present in all his research. An examination of this sort shows that to his way of working "taking a glance at the rich surrounding complexity" we can add the persistence of certain central themes that appear from the beginning, and even though some seem to be left aside for a while, they later reappear. This is what happens, for instance, with probability, the subject of his book published in 2001, *An Introduction to Probability and Inductive Logic*, or with his book of 2004, *Why is There Philosophy of Mathematics at All?"*, which according to Hacking himself is the culmination of his project on mathematics and the fulfilment of his unpublished doctoral thesis of 1960 (Vagelli 2014: 245).

Among the proposed nodes there is a series of relations that can be clearly identified. The style of scientific reasoning or of thinking & doing is illustrated by the probabilistic and statistical style, but it is also an attempt to solve and generalize questions that emerge from the study of those exemplars. For instance, trying to show how after the appearance of the probabilistic style new concepts became possible—such as population—new techniques as the representative sample, and new authorities.[14]

In its turn, the style, given its ability of introducing new objects and kinds, is what offers the conditions of possibility for the emergence, for instance, of new kinds of people, and in this sense, for the making up of people to take place.

This last notion, in its turn, not only appears in the context of a style, but it is the result of his incursion in the subject of probability and statistics, ascertaining that statistics make up every so often new kinds, and in this sense, new people.

The bureaucratic construction of kinds underscores, in its turn, the problems of classification and nomenclature that Hacking develops in his work on kinds—natural and human—and in general terms on doing and intervention.

Moreover, there is also a strong relation between the nodes style of reasoning and experimentation and scientific realism, insofar as the style constitutes the space not

[14] It must be borne in mind that, even though the style of scientific reasoning as a condition of possibility is a prerequisite for the formation of concepts, institutions, etc., it is a necessary but not sufficient reason for the emergence of, in this case for instance, national statistics organisms.

only for the presentation of the realism-antirealism debates due to its capacity for introducing new proper objects and the classifications corresponding to these objects, but also the space for experimentation and the subsequent creation of phenomena.

This analysis also underscores that the style of scientific reasoning can be considered a basal node, since all the others are somehow related to it.

In my view, the thought of Michel Foucault is not just a node in Hacking's work, but rather an underlying thread that runs across already since his reading of *Madness and Civilization: a History of Insanity in the Age of Reason* (1964), even when *The Order of Things* was the work that consolidated the change in his outlook as an analytic philosopher, who until that moment did not consider that the context might be relevant for philosophy. It is the idea that there are historical and situated conditions of possibility that allow for the emergence of scientific objects and concepts, inherited from Foucault but adapted to his own interests, what underlies the whole work of Hacking and achieves the weaving of the network. In the following chapters I will develop this idea.

References

Álvarez Rodríguez, A. (2002). Entrevista con Ian Hacking. *Cuaderno de Materiales*, 17. http://www.filosofia.net/materiales/num/num17/Hacking.htm

Foucault, M. (1964). *Madness and civilization: a history of insanity in the age of reason*. New York: Pantheon.

Foucault, M. (1969). *L'archéologie du savoir*. Paris: Gallimard.

Foucault, M. (1972). *Histoire de la folie à l'âge classique*. Paris: Gallimard.

Goodman, N. (1983). *Fact, fiction and Forecast*. (4th ed.). Cambridge: Harvard University.

Hacking, I. (1965). *Logic of statistical inference*. Cambridge: Cambridge University.

Hacking, I. (1967). Possibility. *Philosophical Review, 76*(2), 143–168.

Hacking, I. (1975a). *The emergence of probability*. Cambridge: Cambridge University.

Hacking, I. (1975b). All kinds of possibility. *Philosophical Review, 84*(3), 321–337.

Hacking, I. (1981). How should we do the history of statistics? In G. Burchell, C. Gordon, & P. Miller (Eds.) (1991), *The Foucault effect. Studies in governmentality* (pp. 181–195). Chicago: Chicago University.

Hacking, I. (1982). Language, truth and reason. In M. Hollis, & S. Lukes (Eds.) (1982), *Rationality and relativism* (pp. 48–66). Oxford: Blackwell.

Hacking, I. (1983). *Representing and intervening*. Cambridge: Cambridge University.

Hacking, I. (1984). Five Parables. In I. Hacking (2002a). *Historical ontology* (pp. 27–50). London: Harvard University.

Hacking, I. (1986). Making up people. In I. Hacking (2002), *Historical ontology* (pp. 99–114). London: Harvard University.

Hacking, I. (1988). The participant irrealist at large in the laboratory. *British Journal for the Philosophy of Science, 39*, 277–294.

Hacking, I. (1989). Extragalactic reality: The case of gravitational lensing. *Philosophy of Science, 56*, 557–581.

Hacking, I. (1990a). *The taming of chance*. Cambridge: Cambridge University.

Hacking, I. (1990b). Natural kinds. In R. B. Barrett, & R. F. Gibson (1990), *Perspectives on Quine* (pp. 129–141). Oxford: Blackwell.

Hacking, I. (1991). A tradition of natural kinds. *Philosophical Studies, 61*(1–2), 109–126.
Hacking, I. (1992a). The self-vindication of the laboratory sciences. In A. Pickering, *Science as practice and culture* (pp. 29–64). Chicago: Chicago University.
Hacking, I. (1992b). "Style" for historians and philosophers. *Studies in History and Philosophy of Science, 23*, 1–20.
Hacking, I. (1992c). Statistical language, statistical truth and statistical reason: The self-authentification of a style of scientific reasoning. In E. M. Mullin (Ed.), *The social dimensions of science* (pp. 130–157). Notre Dame: University of Notre Dame.
Hacking, I. (1993a). Working in a new world: The taxonomic solution. In P. Horwich (Ed.), *World changes. Thomas Kuhn and the nature of science* (pp. 275–309). Cambridge: MIT.
Hacking, I. (1993b). *Le plus pur nominalisme. L'énigme de Goodman: "vleu" et usages de "vleu"*. Combas: L'Éclat.
Hacking, I. (1994). Styles of scientific thinking or reasoning: A new analytical tool for historians and philosophers of the sciences. In K. Gavroglu et al. (Eds.), *Trends in the historiography of science* (pp. 31–48). Dordrecht: Kluwer.
Hacking, I. (1995a). *Rewriting the soul. Multiple personality and the sciences of memory*. Princeton: Princeton University.
Hacking, I. (1995b). The looping effects of human kinds. In D. Sperber, D. Premack, & A. J. Premack (Eds.), *Causal cognition: A multi-disciplinary debate* (pp. 351–383). New York: Oxford University.
Hacking, I. (1998). *Mad travelers. Reflections on the reality of transient mental illnesses*. Charlottesville: University of Virginia.
Hacking, I. (1999a). *The social construction of what?* Cambridge: Harvard University.
Hacking, I. (1999b). Historical ontology. In I. Hacking (2002a), *Historical ontology* (pp. 1–26). London: Harvard University.
Hacking, I. (2001). *An introduction to probability and inductive logic*. Cambridge: Cambridge University.
Hacking, I. (2002a). *Historical ontology*. London: Harvard University.
Hacking, I. (2002b). *L'émergence de la probabilité*. Paris: Seuil.
Hacking, I. (2002c). How 'natural' are 'kinds' of sexual orientation? *Law and Philosophy, 21*(3), 335–347.
Hacking, I. (2007). Natural kinds: Rosy Dawn, scholastic twilight. In *Royal Institute of Philosophy supplement* (pp. 203–239). Cambridge: Cambridge University.
Hacking, I. (2009a). *Scientific reason*. Taipei: National Taiwan University.
Hacking, I. (2009b). Discours de réception lors de la cérémonie de remise du Prix Holberg, 25 novembre 2009. *La lettre du Collège de France*, 27. http://lettre-cdf.revues.org/356
Hacking, I. (2010, April 21). *Lecture I. Methods, objects, and truth*. [Unpublished]. México: UNAM.
Hacking, I. (2011a). Wittgenstein, necessity and the application of mathematics. *South African Journal of Philosophy, 30*(2), 155–167.
Hacking, I. (2011b). Why is there philosophy of mathematics AT ALL? *South African Journal of Philosophy, 30*(4), 1–15.
Hacking, I. (2014). *Why is there philosophy of mathematics at all?* Cambridge: Cambridge University.
Martínez, M. L. (2009). *Realismo científico y verdad como correspondencia; estado de la cuestión*. Montevideo: Facultad Humanidades y Ciencias de la Educación.
Vagelli, M. (2014). Ian Hacking. The philosopher of the present. *Iride, 27*(72), 239–269.
Williams, B. (2006). *Verdad y veracidad. Una aproximación genealógica*. Barcelona: Tusquets.

Chapter 2
Styles of Scientific Thinking & Doing. A Genealogy of Scientific Reason

My discussion of styles appropriates much Foucault, as I understand him.

Ian Hacking (1992c: 137)

Abstract In this chapter, *Styles of scientific thinking & doing. A genealogy of scientific reason*, I work on the "style of scientific reasoning", later called "style of scientific thinking & doing". I deal with this node in the first place because I visualize it as basal, both because it is a condition of possibility for the emergence of objects and concepts that appear in other nodes, and because one way or another all the others are related to it. At the beginning of the chapter I deal with the antecedents, differences and convergences that Ian Hacking's notion of style of scientific reasoning has with meta-concepts proposed by other philosophers and historians of science. I show how already in 1979, in an article where Hacking offers his interpretation of Michel Foucault's thinking, he discusses some of what will later be characteristic of his own notion of style. I deal in particular with the notion of style of scientific thinking of Alistair Crombie, which Hacking considers an immediate antecedent of his notion of style of scientific reasoning. I continue by going deeper into the notion of style of scientific reasoning by means of an analysis of its particular features, as well as their relation to notions proposed by Foucault.

Keywords Styles of scientific thinking & doing · Michel Foucault · Alistair Crombie · Antecedents of Ian Hacking's style of scientific reasoning · Features of style of scientific reasoning

M. L. Martínez Rodríguez, *Texture in the Work of Ian Hacking*, Synthese Library 435, https://doi.org/10.1007/978-3-030-64785-8_2

In 2010 (April 21) Hacking mentions his list of interests regarding the style of scientific thinking & doing: genealogy, history of the present, innate human abilities, cognitive history and anthropology. His aim is to develop on this basis a genealogy of scientific reason, to find out how we investigate. To analyze the different styles of scientific thinking & doing, the various general methods of scientific work that can be recognized from antiquity to the present.

2.1 Antecedents of Style of Reasoning or Style of Scientific Thinking & Doing

Even though Hacking situates the beginning of his project on styles in 1982 with the article "Language, Truth and Reason", I have claimed that, in spite of the fact that in *The Emergence of Probability* the notion of style of scientific reasoning does not appear explicitly, this text could be considered as belonging to this line of research. Now I will add that 4 years after this text and 3 years before Hacking's remark, in his article "Michel Foucault's Immature Science" (1979), he claims that the first chapters (2 to 6) of the aforementioned book are perhaps the only detailed study—in English—of a change of style in rationality. These chapters, he also claims (1979: 41), owe much to *The Order of Things*. Neither will Hacking in this article, which deals with the Foucauldian conception of the life sciences, work and language, and which Hacking calls –problematically—immature,[1] mention the notion of style. However, in presenting his interpretation of the thought of the French philosopher in this area, Hacking advances in some of which will later be features of his proposal of the style of scientific reasoning.

To answer the imaginary question of comparing and contrasting Foucault's archaeology with the American theory of knowledge of that time, Hacking elaborates, in a way that is at the same time prosaic and abstract, a list of six hypothesis with which, he says, Foucault starts his enterprise:

[1] Hacking uses the expression immature sciences to refer to disciplines "*whose foci are 'life, labor, and language*'" (1979: 40), which Foucault deals with in *The Order of Things*. Hacking does not explain, at least in this article, why he gives them this name, but he alludes to Putnam's use of the expression in *Meaning and the Moral Sciences* (1978), which leads me to think that the distinction between mature and immature applies to sciences whose theoretical terms refer and sciences whose theoretical terms do not refer. He also develops the respective proposal of Thomas S. Kuhn in *The Structure of Scientific Revolutions* (1962). Hacking's use of the expression immature science is, to say the least, problematic, insofar as even admitting that none of the two above mentioned philosophers of science have said much about immaturity, he uses the distinction to distinguish between Foucault's work and that of the other two philosophers. It is problematic also because Foucault has specifically vindicated that the human sciences have a different kind of rationality from the natural sciences, not due to immaturity in its development or methodology, but rather due to a fundamentally different organization of knowledge, and because he has claimed that these sciences are not in a relation of teleological subordination respect to the orthogenesis of sciences.

1. The systems of thought in immature sciences exhibit defined laws and regularities for which the hypothetical-deductive model is irrelevant.[2]
2. These regularities determine systems of possibility.
   This is what both Foucault and Hacking call positivity. What counts as reason, argument or evidence, is part of a system of thought, so that the modes of rationality are temporal and not eternal. What is conceived as true or false, what counts as a basis to sustain or refute which argument or data are relevant, is delimited by these regularities.
3. Immature sciences are not modelled on definitive realizations and must be studied through the anonymous mass of materials, rather than through a few spectacular achievements. Competing bodies of beliefs have the same underlying formation rules. The great positive realizations within a system of thought merely "fill up" or elaborate certain pre-established uniformities.
4. The regularities that determine a system of possibilities are not a conscious part of it and they might not be articulated within it.

Hacking remarks that Foucault has used concepts such as *savoir* in relation to this, understanding it as the elements that must have been formed by a discursive practice so that eventually a scientific discourse might be constituted (Foucault 1969: 237, 1972: 182). The set of elements formed on the basis of a single positivity, in the field of a unitary discursive formation.

He refers to the discursive conditions necessary for the development of *connaissance*, conditions that are necessary in order to know certain objects and to formulate certain statements in a particular period. It provides the objects, kinds of cognitive authority, concepts and subjects necessary for a body of scientific knowledge. In other words, it provides the pre-knowledge or necessary knowledge for the constitution of a science, even if it is not necessarily destined to give it place. It is defined by the possibilities of use and appropriation that a discourse offers.

It is what must have been said –or what must be said—so that a discourse might come into existence which, if necessary, would answer certain experimental or formal criteria of scientificity (Foucault 1969: 238, 1972: 182).

5. The surface of a system of thought is what is actually said. Neither intentions nor meanings play a central role in the analysis.
6. There are acute discontinuities in the systems of thought, followed by periods of stability.

What Hacking will then mention as his main innovation with respect to Crombie's list of styles of scientific thinking is certainly his argument that the styles of scientific reasoning, in their long-term existence, are interrupted by what he calls crystallizations (2009, 11). Revolutions are of interest since they are the beginning of the

[2] It is irrelevant because the irregularities he mentions are not a specific theory and the consequent model that in a mature science with well-articulated postulates leads, almost by deduction, to a rich sample of verifiable hypotheses. The sciences discussed here are unrelated to a body of theses but to a system of possibilities. Hacking (1979).

regularities that explain normal science—in Thomas S. Kuhn's words—but it is normality that is actually interesting to understand how the systems of possibility can ground how we think. Foucault's hypotheses contribute to understanding that it is not particular theories that are incommensurable, but that incommensurability is in the bodies of discourse, in the systems of possibility. What the system of possibility of a particular theory has as the grounds for truth or falsehood does not coincide with what in other moments or in other bodies of discourse is understood as such. This is not due to translation, which in this case is irrelevant, but to the incommensurability of the systems of possibility. What is needed is to share the same system of possibilities.

As will become clearer below, there appear in this brief review of Foucault's ideas antecedents of what will later characterize Hacking's notion of style of scientific reasoning. Among them are: positivity as an attribute of the style of reasoning; its capacity to determine what can be said, what counts as proof, as true and false; the existence of ruptures or discontinuities in the development of the styles and the idea of incommensurability, albeit not an incommensurability in the Kuhnian style.

Some years later, in articles such as "Language, Truth and Reason" and "Five Parables", Hacking refers to the debts that his notion of style of reasoning owes Alistair Crombie's idea of style of thinking—developed in *Styles of Scientific Thinking in the European Tradition* (1994)[3]—and with the notion of style of thinking of Ludwik Fleck's—developed in *Entstehung und Entwicklung einer wissenschaftlichen Tatsache* (1935). But it is only in articles of 1992 that Hacking explains more and better the relations between his notion of style of reasoning and these preceding ideas.

In "'Style' for Historians and Philosophers", he claims he adapted to metaphysics and epistemology A. Crombie's idea of styles of scientific thinking in the European tradition, changing its name to styles of scientific reasoning because he considered that "*reasoning was a slightly less mentalistic gerund than thinking*" (Hacking 2010, April 21: 9), "*avec davantage de connotations liées à la recherche et à la découverte*" (Hacking 2003: 543). He also mentions that even though Fleck's "style of thinking" is restricted to a discipline or field, his notion of style of reasoning shares with it being the possession of a perdurable social unity, a collective thought. Finally, there are also in this text several references to Foucault illustrated by what Hacking expresses about the French philosopher when talking about the positivity of style: his "*influence on my idea of styles of reasoning is more profound than that of Comte or Crombie*" (Hacking 1992c: 12).

In that same year, in an article entitled "Statistical Language, Statistical Truth and Statistical Reason: the Self-Authentification of a Style of Scientific Reasoning",

[3] Crombie, A.C. (1994) gathers in three volumes Crombie's lifelong work. Here the author outlines the trajectory of six styles of thinking, from the Mediterranean world of Antiquity to modern Europe. Hacking heard Crombie talk about styles of scientific thinking in a conference in Pisa, in 1978, and according to his own words, "*that was a sort of turning point, a kind of "click", because you can see from my background, with my interests in Foucault and history, how this idea could appeal to me*" (Vagelli 2014: 255).

Hacking contrasts his idea of style of reasoning with the most familiar meta-concepts of the last third of the twentieth century. Among them, there is Foucault's idea of discursive formation.[4] The relation and comparison of this thought with the ideas of the French philosopher is reiterated and once more Hacking leaves no doubts about his debt with him as he claims that his discussion of styles appropriates much of Foucault, as he understands the French philosopher (1992c: 137).

In 2009, in *Scientific Reason*, he mentions a series of shortcomings of the use of the word *style*[5] -an expression originally chosen because it suggested the idea that the forms of "*finding out*" are intrinsically plural. He continues saying that he has abandoned the expression "styles of scientific *reasoning*"—even though he uses repeatedly in the text in question—and he has returned to Crombie's "style of scientific *thinking*". With an explanation reminiscent of Thomas Kuhn when he claims that the concept of paradigm is interpreted in as many different ways as there are interpreters, Hacking remarks that in abandoning the notion of style of reasoning he leaves it for others to use it in their own senses. He will speak from now on about styles of scientific thinking and, using Ockham's maxim, he will not extend the list of style beyond what is necessary in order to avoid the tedious debate on how to define them.

A year later, in a series of conferences given in Mexico, he claims that:

> Science is as much a matter of activity as of thought. Since I want to emphasize action and intervention, I speak of styles of scientific thinking & doing. I don't like the word 'style', but it has become entrenched in my expositions. To avoid confusion, I shall stick with it. (Hacking 2010, April 21: 3)[6]

Finally, in 2014 (142), in a retrospective look at his use of the word *style, Hacking* claims he has done it in very different senses, as a general concept that gathers the more habitual forms in which a subject is thought about. Now, he claims, he speaks indistinctively about styles of scientific thinking & doing, or genres of inquiry, or simply ways of finding out. The ways of finding out is what have enabled human beings to dominate the planet and they are grounded in human capacities which have evolved in evolutionary time.

---

[4] In *The Archaeology of Knowledge and the Discourse of Language* Foucault characterizes discursive formation as "[…] *the discursive formation is the general enunciative system that governs a group of verbal performances—a system that is not alone in governing it, since it also obeys, and in accordance with its other dimensions, logical, linguistic, and psychological systems.*" (1969: 152, 1972: 116). A discursive formation "[…] *determines a regularity proper to temporal processes; it presents the principle of articulation between a series of discursive events and other series of events, transformations, mutations, and processes.*" (1969: 98–99, 1972: 74).

[5] The notion of style is unsatisfactory both because it is too volatile and because it has acquired numerous connotations with its use in different contexts, not only scientific but also in aesthetics and many other fields.

[6] From here onwards we will try to respect Hacking's use of these terms, referring to the style of scientific reasoning or the style of scientific thinking & doing in the same sense and on the same occasions in which he does. However, it must be clear from this point that in using one or the other locution we are referring to the same notion.

I consider the expression styles of scientific thinking & doing to represent more appropriately than the concept of scientific reasoning the idea Hacking has sustained from the beginning, namely that styles are not just related to thinking but to doing and manipulating in scientific practice. We think with our hands, our whole body (Ibid., xiv).[7]

2.2 Metaphysics, Microsociology and Anthropology

Hacking's interest is to study how we find out, to make a genealogy of scientific reason, in the sense of "how we found out how to find out" (Hacking 2010, April 21: 3).

He is concerned by the fundamentally different ways in which civilizations have discovered how to change the universe. His idea is basically metaphysical and philosophical: how human being have learned to conquer the world. It is a very Baconian idea: knowledge is power.

He is interested in analyzing the basic, distinct and complementary styles of scientific thinking & doing, the various general methods of engaging in scientific work that can be recognized in ancient times, which solidified over the course of centuries, and which continue to be practiced in full vigor today. Mathematical proof, developed for the first time in geometry, was the first to be systematically deployed in ways we still today recognize as ours. It is a history, certainly an anachronistic one– and Hacking knows it—of how the present moment has been arrived at (Hacking 2010, April 21: 4).

In his article "'Style' for Historians and Philosophers", Hacking presents the style of reasoning as "*a new analytical tool that can be used by historians and by philosophers for different purposes*" (1992b: 1). He proposes it as something at once social and metaphysical, capable of bridging the gap between social studies of knowledge and philosophic-metaphysical conceptions of truth, existence, logic, meaning, etc. He explains he prefers to use the term reasoning instead of thinking to mean not just argumentation but also manipulation, not just private but also public, not just speaking but also arguing and showing (Ibid: 3).

[7] Even though it might not seem a descriptive style such as the statistic, it clearly shows, according to Hacking, how the development and growth of a style of reasoning are a question not just of thinking but of doing. It may seem surprising that the statistical descriptive style, which results from a mere enumeration, deserves the name of style of scientific thinking & doing. However, the cases of descriptive census are extremely complex; the census must be organized, the surveyors chose, a code invented, as well as tests to find the surveyors that falsify numbers, etc. All these activities form part of the reasoning. The data is not passive, it is recorded, collected, moved, ordered, systematized.

Likewise, mathematics has the function of *making* things. We think, talk, gesticulate, make diagrams. Hacking considers mathematics as something done in a material way. (Hacking 2014: xiv)

Whereas Hacking takes as his basis the list of styles of thinking of this Australian historian of science:

1. The simple method[8] of postulation exemplified by Greek mathematical sciences;
2. The deployment of experiment both to control postulation and to explore by observation and measurement;
3. The hypothetical construction of analogical models;
4. The ordering of variety by comparison and taxonomy;
5. The statistical analysis of regularities of population and the calculus of probabilities;
6. The historical derivation of genetic development (Ibid:4) (Crombie 1988: 10–12),[9]

he does not agree with Crombie's project and undertakes his own (Hacking 2003: 543), proposing a provisional list of styles of scientific reasoning similar to Crombie's:

1. Mathematics: (1a) geometric style, (1b) combinatorial style

Mathematics do not constitute a unified whole but rather a cluster, that can be subdivided at least in the two aforementioned sub-styles. The geometric one, developed among the Greeks, introduces a new way of telling the truth: mathematical proof, departing from which it not only possible to understand, for instance, that a theorem is true, but also why it is so. On its part, the combinatorial style belonged to the Islamic, Persian and Hindu world, and introduces a new way of telling the truth about quantities.

2. Laboratory style

The most powerful style of reasoning, which not only has made the modern world possible but also allowed its permanent change until the present, is the laboratory style. Hacking presents it as an example of composition of two classical styles, Crombie's styles 2 and 3, and characterizes it by the construction of devices to produce phenomena that can be true or false according to hypothetical models. He adds, besides, that even though it is a combination of the aforementioned styles it does not replace them; on the contrary, there are fields of specialization in which experimentation or modelling continue unchanged.

[8] In the beginning, Crombie called them both methods and styles, but he gradually gravitated to the second concept.

[9] Crombie divides the list in two groups, 1–3 and 4–6. Hacking, for his part, points out two differences between both groups. The first one is related with the power of the styles in the first group, which have changed the face of the Earth and allowed us to become its parasites. The second has to do with truth. Truth is an objective of the mathematician, of the experimenter, of the theoretician, but not, in the same terms, of the taxonomist. Classifications are not seen frequently as true or false, but rather as suitable, informative, explicative. This is not to say that the words truth or falsity are not used in taxonomic discourse but taxonomic structures as a whole are not judged as true or false but as better or worse. (Hacking 2010, April 26: 5).

Its beginnings can be traced back, according to Hacking, to the time when Robert Boyle constructed the vacuum pump. This scientist and his device are, for him, the emblem of the crystallization of the laboratory style. Its core is the construction of instruments designed not only to examine the world but also to intervene in it, for the creation of new phenomena.

The laboratory is an institution where work is carried out with phenomena that rarely happen in a pure state before they are constructed in it under surveillance and control. It is at the same time a public and a private space. Public because any work carried out in the laboratory can be reproduced by any other researcher with the right tools and checked by anyone who is a good observer. But also private because only a self-selected few can understand what is happening in it.

3. Galilean style (of hypothetical modelling)

Simulation plays a central role in this style, that is to say, the representation of aspects of reality by means of a process of hypothetical construction of analogical models, grounded on structures frequently inaccessible observationally. Its central figure was Galileo, who used mathematical models to understand the phenomena of nature.

4. Taxonomic style

Despite claiming that Linnaeus could serve as an emblem of what he will call crystallization of the taxonomic style due to the fact that at that moment a radical change took place from which there was no return, Hacking claims he is not satisfied with the statement that the taxonomic style crystallized in Linnaeus' time. What happened was, instead, the acquisition of a paradigm of its own. Taxonomic thought acquired a model of how to advance using a hierarchical structure of taxa. It was a way of simplifying the classification of organisms and facilitate the recall of their names, that was slightly modified and became canonical during the eighteenth century. However, this paradigm has many of the effects of a crystallization; for instance, the emergence of new objects.

5. (5a) Probabilistic style; (5b) Statistical style

The keyword in these styles is "population", an innovation engendered by the statistical style. They study the regularities of populations ordered in space and time. "Statistics" and "population" are inextricable concepts; they evolved together, each one at its own pace. Statistics was introduced to account for some conditions of certain European states, but gradually became applied to more and more varied kinds of groups, until the word "population" became a significant abstract term of any specific collection of objects.

6. Historical-genetic style

According to Hacking, this style starts with the myths on the origin of things, it develops as a style of reasoning, invoking efficient causes and causal processes, and finally turns into explanations that can take a purely deductive-nomological form. This is the case because the historical style makes use of a particular form of reasoning which more than in the establishment of truths, emphasizes the discovery of explanations. Hacking situates the beginnings of this style in scientists such as Buffon, which had a historical vision of nature, interested in understanding the change both from a mathematical and an experimental perspective.

Hacking disagrees with Crombie's list of styles of thinking due to three fundamental reasons:

1. His exposition of the first three styles is exhausted by the end of the seventeenth century. Only the last style is developed in the nineteenth century with Darwin as its main figure. Hacking, for his part, remarks that the history he wants is the history of the present (Hacking 1992b: 5). This idea taken from Foucault implies that we recognize and distinguish historical objects to illuminate our own situations. This is why Hacking can modify Crombie's list, not to revise his history but to consider it from the present. And in the article "Styles of Scientific Thinking or Reasoning" (1994: 35), complementing what he had already said about Foucault, adds that he can be as 'archaeological' as he wishes, but the selection of a historical object is always made with a present aim in sight.
2. Crombie's enumeration is a historical progression. Each style starts after its predecessor on the list and concludes closer to the present. Hacking is moved by the fact that the six styles remain alive at present, even though what is important now can be different from what used to be important before.
3. Crombie's list is not exhaustive. He transcribes what he found of permanent and central in the formative period of the western vision. However, there are also styles to register, both earlier and outside the west, and not identified as a mere anticipation of the first style in Crombie's list. Moreover, new styles can also be developed, subsequent to the classical ones mentioned by the Australian historian, and others can also appear as the result of the composition of two classi cal ones.

Crombie has a history of styles. Hacking adds metaphysics, microsociology of the origins and philosophical anthropology. Metaphysics because, as we shall see, each style establishes new statements, objects, kinds of entities, classes, kinds of classification, explanations, accuracy criteria, laws. Each style carries a multitude of new things that were not said earlier, statements that were not uttered and furthermore, that could not be uttered. It introduces its own criteria of proof and demonstration and determines the appropriate conditions of truth for the domain in which it can be applied. This leads Hacking to radical theses on truth and objectivity and about the reality of scientific objects themselves. A style of scientific reasoning is more than a group of techniques to propose new kinds of facts in our knowledge. Styles are framing concepts.

The microsociological aspect—that is to say, the immediate specificities of the agents and the institutions—is due to the fact that although Hacking's list—like Crombie's—is internal, he believes that an important lesson of microsociology about the circumstances under which a style emerges must be taken into account, since there is nothing that makes the emergence of a style of reasoning necessary (Hacking 1995). Hacking adopts this idea from Ludwik Fleck, despite maintaining some differences with his proposal. For instance, Wassermann's test for syphilis – which Fleck takes as a reference—is an issue, at the most, of decades, whereas the scientific styles Hacking is thinking about are more long-term. Fleck speaks specifically of a form of thinking, of research, in force within a specific collective, in a defined time, which evolves, mutates and dies in a brief lapse of time. Hacking's style, for their part, are less local and more durable, partly because they are constructed on the basis of fundamental cognitive capacities. They are not extinguished by refutation but abandoned.

Finally, Hacking adds philosophical anthropology, in the sense that styles become autonomous from their origins. There are physical and psychological facts that lead to the autonomization of a style of reasoning. Such facts do not justify the style in the sense of showing why it leads to the truth, but are the background that allows for the existence of a self-authenticating style. Innate human abilities and the development of social institutions constitute an intellectual orientation through which scientific reason can be understood. The way self-authentification techniques work depends on facts related to people and their place in nature. But the detailed interest of self-authentification requires paying attention to the peculiar techniques of each style. Each style is based on innate human capacities, which are discovered, exploited and developed in specific historical situations. Thus styles are the product of cognition and culture, of interaction between, on the one hand, unique human endowments related to our evolutionary heritage, and on the other, specific historical events and developments. People have learned how to use their abilities. As human culture develops, we discover how to use these abilities in a completely new way, which is not completely predetermined biologically. There is a "wrapping" of sorts that restricts our possibilities, but what we do within it has to do with human development. Cognitive abilities are but the framework within which we learn to do all sorts of things. We learn to find out; this is not something predetermined. Our framework provides us with a series of capacities, which are possible to develop; however, this development depends on how they interact with the world and what happens with it afterwards. In this way the interaction of culture and nature produce a looping effect of sorts.

Given the innate cognitive elements, moreover, it follows that, if a style of reasoning was developed first in a single historical culture, it can also be acquired by people in other cultures.

Hacking claims that his main innovation is organizational, since many of the techniques he describes are very well known. The most unusual feature of styles is their quality of being capable of self-authentification. This self-authentification and/or self-justification[10] is the cornerstone for objectivity, and in the laboratory, for

[10] Hacking uses the expression self-authentification to mean the way in which a style of reasoning generates the conditions of truth for their own propositions. He understands by self-justification the way in which ideas, things and marks mutually adjust in the laboratory. Thus self-authentification is a logical concept, whereas self-justification is a material concept (Hacking 1992a: 51, footnote 2).

reproducibility. Moreover, the introduction of new kinds of objects and new ways of verifying statements about them are a source of stability for the sciences.

2.3 Anonymous, Autonomous and Common to Several Sciences

The style of scientific reasoning is, as I have stated, a perdurable and impersonal social unit; it is the intellectual preparation or availability of a particular form of seeing and acting. It is, as Foucault remarks, apropos discursive formations, an anonymous and autonomous system, which is not constituted by the beliefs of a person or a school.

Regarding discursive formations, Foucault states that it is not by resorting to a transcendental subject or to a psychological subjectivity that the regime of the statements of a discursive formation should be defined. Enunciative modalities are defined on the basis of a law that operates behind the diverse statements, a law that avoids referring to objects or subjects.

Statements appear and circulate in a field that presupposes forms of order and succession, forms of coexistence (acceptance or rejection) and procedures of intervention. It is what Foucault calls the associated field of the statement. Thus he appeals to a level of analysis that accounts for continuities, small changes and radical reordering of the concepts without resorting to an immanent rationality, that is to say, without claiming that a theory is replaced by another one because the new one is superior according to some rational general principle. He chooses to remain on the level of the systems of discursive practices which, he claims, are autonomous and governed by rules; on the level of the rules actually applied in discourse, which have their place not in the mentality or consciousness of individuals, but in discourse itself; they are imposed, therefore, according to a uniform anonymity of sorts, to all the individuals that are willing to speak in that discursive field (Foucault 1969: 83–84, 1972: 63).

With regards to anonymity, the style of reasoning also has an antecedent in Fleck, since it is, as his style of thinking, a collective thought. However, it does not determine, as the Fleckean concept, a specific content or science, but rather methods that can be deployed in any science. A style is common to several of them, and in its turn, the same science uses many styles. Evolutionary biology, for instance, uses the mathematical style; measurements and experimental exploration; hypothetical models and analogy; the taxonomic style; probability and statistics, and it is the most viable example of a historical-genetic science. The styles of scientific reasoning are different forms of finding out, of participating in the different kinds of scientific activities. In this sense, both Hacking and Foucault propose conceptual structure of a wider range than those proposed by Fleck and other philosophers of science such as Thomas S. Kuhn. Discursive formations, even though they can be closely related to the sciences, are not identified with them and cannot be identified as prototypes of future science. Let us remember that the relation between discursive formations and the sciences is based on the special sense that Foucault gives to the distinction between *connaissance* and *savoir*, understanding the former as any particular body of knowledge and the latter as the discursive conditions necessary for the development of the former one. A discipline is the *locus* of *connaissance* whereas a discursive formation is the *locus* of *savoir.*

Even though the style of scientific reasoning is not identified nor is it exclusive of a particular science or scientific community but rather runs through and is shared by several of them, in my view and unlike the discursive formation, neither does it shape the former, whence sciences emerge. Neither does a style belong to a certain epoch, as does the episteme[11] and the discursive formation. In this sense it is not identifiable neither with *savoir* nor with *connaissance*. Even though it adopts some features, for instance, fundamentally from the concept of *savoir*, such as being the space in which the subject can take a stand to speak about the object they deal with in their discourse, it is also the field of coordination and subordination where concepts appear, are defined, applied, and transformed.

Despite the familiarity of the style of reasoning with Foucault's proposal, despite the fact that much of what Hacking wrote about the style of reasoning can be found in one or more Foucauldian notions, the former is not identified, strictly and *in toto*, with none of the latter. A style is not, like Foucauldian episteme, for instance, the set of relation that link, in a given epoch, the discursive practices that give place to some epistemological figures, some sciences, eventually some formalized systems (Foucault 1969: 250, 1972: 191).

A style does not give place to the sciences, but rather it is shared by many of them. It does not belong to an epoch but rather runs across it. Neither is it, like episteme, a single one in a given culture and time. On the contrary, according to Hacking, several styles coexist at the same moment and in the same culture, and they are even used by the same science. In this sense, Hacking remarks that styles must not be conceived as isolated entities, but as compatible and complementary, interacting

[11] Episteme defines the field of analysis of archaeology. Foucault moved from a monolithic conception of episteme in *The Order of Things* to a more open conception in *The Archaeology of Knowledge and the Discourse of Language.* In the first work he tells us that: "*Thus, between the already 'encoded' eye and reflexive knowledge there is a middle region which liberates order itself: it is here that it appears, according to the culture and the age in question, continuous and graduated or discontinuous and piecemeal, linked to space or constituted anew at each instant by the driving force of time, related to a series of variables or defined by separate systems of coherences, composed of resemblances which are either successive or corresponding, organized around increasing differences, etc.*" (Foucault 1966: 12, 2005: xii–xiii). In *The Archaeology of Knowledge and the Discourse of Language* he claims that by episteme it is understood "*the total set of relations that unite, at a given period, the discursive practices that give rise to epistemological figures, sciences, and possibly formalized systems; the way in which, in each of these discursive formations, the transitions to epistemologization, scientificity, and formalization* [...] *The episteme is not a form of knowledge (connaissance) or type of rationality which, crossing the boundaries of the most varied sciences, manifests the sovereign unity of a subject, a spirit, or a period; it is the totality of relations that can be discovered, for a given period, between the sciences when one analyses them at the level of discursive regularities*" (1969: 250, 1972: 191).

between them. They are not so separated, but are rather instruments that fuse with each other, Hacking claims, even though he does not provide detailed examples of how this integration of styles would take place.

Whereas episteme is not a kind of knowledge or a kind of rationality that runs through the most diverse sciences (Foucault 1969: 250, 1972: 191), a style could actually be considered to be in some sense a kind of rationality that runs through several disciplines.

If the episteme is the specific order of knowledge, the configuration, the arrangement that knowledge takes in a specific epoch and that confers it a positivity as knowledge, it is not easy to identify which is the specific, clearly delimited order that might correspond to a style. It is not the knowledge of an epoch, it is not a science, they are ways of thinking and doing that, like Collingwood's absolute presuppositions,[12] function as basic assumptions for a certain collective of people who do not necessarily need to share amongst them something other than the style. That is to say, there could be scientists that share an epoch but not a style and, conversely, there could be scientists that share the style but not the epoch. There might be scientists within the same science that use different styles, whereas scientists of different sciences could share the style. If the episteme makes it possible to grasp the play of coercions and limitations imposed on discourse at a given time, we can say of style that it makes it possible to grasp them for the discourses that correspond to each style, the probabilistic discourse, the mathematical, taxonomic, laboratory sciences discourses.

Style seems to be, regarding its range but not its location, something intermediate between the episteme or the discursive formation and the sciences.

López Beltrán (1997: 144) has remarked that Hacking does not seem to feel comfortable with such long-term and wide-ranging generalizations as Foucault's. On the other hand, there is a sense in which Hacking's styles have been questioned for being too broad. Bueno (2012) understands Hackinian styles as models of inferential relations used to select, interpret and support evidence for certain results. He has warned how the breath in the nature and reach of these can turn them irrelevant

[12] Robin Collingwood's (1940) *Essay on Metaphysics* is a book about metaphysics, understood as the science that deals with the presuppositions that underlie science. Presuppositions are not priorities in time but logical priorities. Every question contains an absolute or relative presupposition. A presupposition is relative if it is posed relatively to a question as its presupposition and relatively to another question as its answer. An absolute presupposition, on its part, is that which is posed relatively to all questions as a presupposition and never as an answer. Absolute presuppositions are not verifiable. The idea of verification does not apply to them since talking about verification of a presupposition assumes that it is a relative presupposition. Presuppositions are not propositions, because, once again, every proposition is the answer to a question and absolute presuppositions are never answers. Absolute presuppositions are not consciously sustained, and neither are conscious the changes produced in them and which lead to abandoning what seemed to be the strongest habits and guidelines. Absolute presuppositions form, at each moment in history, a structure that suffers tension, that can collapse and be replaced by another one. Even though constellations of absolute presuppositions in Collingwood do not have some of the particular features of styles of scientific reasoning, they somehow are comparable to them, insofar as they appear historically and are self-authenticating, albeit in a silent way.

for the understanding of the details of scientific practice. As a consequence, he proposes a narrower unit of analysis: narrow styles of reasoning, which, even though it preserves certain features of generality insofar as it is centered on the inferential role of several aspects of scientific practice, he offers more specific information about the particular domains of investigation, providing a focalized outline that can give sense to the significant aspects of the aforementioned practice. This proposal does not undermine, according to Bueno, the actual idea of style of reasoning nor its utility, but has a remarkable difference with Hacking's style regarding the specification of what propositions are true or false. Narrow styles only formulate what is considered possible within a certain domain. And this can be done independently of any particular commitment with the truth or falsehood of the discourse in question. The exploration of the possible, given the accepted characteristic in the domain of investigation is a central aspect of a narrow style of reasoning.

2.4 The Relation Style-Ontology

In Crombie's words, styles can be "*distinguished by their objects and their methods of reasoning* (Crombie 1994: 83). Objects and methods of reasoning: two words that will provide Hacking a pivot to give a turn to Crombie's particular history and achieve an even more singular philosophy.

Hacking claims that each style introduces their own distinct kinds of scientific objects, that is to say, it is a condition for the emergence of certain objects. He introduces a battery of innovations that includes new kinds of objects, elements of proof, sentences, laws, possibilities, new kinds of classification and explanation. Each style is specific in its own domain only because it introduces its particular objects.

The idea is not that we first have a style of reasoning that then introduces a new kind of objects and a new method of reasoning. In this emergence there are no steps; the style and the elements appear together, they are simultaneous. Styles are constituted by their methods and the kinds of objects with which they deal.

The introduction of a new domain of objects of study generates for each new class of entities an ontological dispute, a debate about what exists, because the new kinds of objects are individualized on the basis of the style itself and are not previously evident. Thus, the ontological debates about the abstract objects of the mathematical style, of unobservable theoretical entities in the style of hypothetical modeling, of taxa in the taxonomic style, are a product of the introduction of new kinds of objects in the course of the emergence, acceptance and use of a new style of reasoning. Ontological debates only take place within their own scientific style. They result from the introduction of objects by styles of reasoning, the fact that we speak about those objects using statements in which the names of the objects serve as grammatical subjects, and that many languages demand an existential presupposition for the terms in the subject position. This undermines the traditional idea of the debate between realism and antirealism, which is frequently presented in general terms. These debates do make sense, according to Hacking, except in the context of a style of reasoning.

Foucault's discursive formation also produces the objects about which it speaks. The regime of existence of the objects of a discourse, according to Foucault, is made by avoiding any anthropological subjection, respecting the level of enunciative analysis; referring them to the set of rules that makes it possible to form them as object of a discourse and thus constitute their conditions of historical emergence. These rules are manifested on three levels (Foucault 1969: 56–58, 1972: 41–42), namely: (1) The surfaces of emergence, that is to say, the points or moments of appearance of an object departing from successive and simultaneous accumulations of practices, statements and discourses that have a certain degree of coherence. Likewise, the style produces its objects. (2) The instances of delimitation, that is to say, the process by means of which an object is isolated and acquires recognition on the part of the different social strata that designate, name or install it: Hacking's experts. (3) The grids of specification, systems according to which the different objects of discourse are separated, opposed, regrouped, classified or emanate from the different objects of discourse. In Hacking, the specific classifications of each style.

According to Foucault, objects do not exist by themselves, they do not come from the empirical and objective substrate of experience, but they are logical configurations that emerge from a specific discursive formation under certain conditions of visibility and intelligibility determined by the perceptive field from where they emerge. The object is not an *ob-jectum*, something placed in front of the subject. Its formation depends on the relations established between emergence surfaces, instances of delimitation and grids of specification, to the extent that, at least regarding its objects, a discursive formation is defined if it is possible to show how any object of the discourse in question finds in it its place and its law of emergence (Foucault 1969: 60–61, 1972: 44).

The aforementioned relations, called discursive relations, do not explain how the object is constituted but why at a certain epoch the space was created for its emergence. Hence the relation between words and things necessarily imposes treating discourses no longer as lexical, linguistic or significant rules, but rather as practices that generate and produce the objects to which they are applied. Both non-discursive and discursive practices play a role in the formation of the objects, but Foucault mainly stresses the crucial role played in this aspect by discursive relations. These relations are not thought of as logical or rhetorical relations between propositions, but rather as the relations between speech acts used in specific contexts to carry out certain actions. Discursive relations are, in a certain sense, at the limit of discourse, "*[...] they offer it objects of which it can speak [...], they determine the group of relations that discourse must establish in order to speak of this or that object, in order to deal with them, analyze them, classify them, explain them, etc.*" (Foucault 1969: 63, 1972: 56).

Hacking, for his part, emphasizes, besides discourse, all those social, economic, etc. conditions that surround the emergence of objects within the style.

2.5 Stabilization Techniques

Each style develops an ensemble of techniques that ensure its stability. The existence of such techniques is the condition for a style to be able to: produce a relatively stable body of knowledge and ensure the openness, creativity, self-correction capacity and continually engender new knowledge and new applications. These are part of the necessary conditions for a style of scientific reasoning.

An overview of each one of the stabilization techniques requires a detailed analysis, specific for each style, as well as a historical illustration. Almost the only thing the stabilization techniques of the different styles have in common is that they fulfil this function. Some techniques are more effective than others. The taxonomic and historical-genetic styles, for instance, have not produced anything like the stability of the laboratory or the mathematical styles. In the case of the laboratory style, its stabilization technique is not a mental or social construction. It is material. It has to do with the relation between thoughts, acts, and technical productions. A theory predicts something that can be observed in a certain time and place. For its part, the statistical style is so stable that it has its own word to refer to its more persistent techniques: "robustness".

Even though the style persists in its individual and peculiar way precisely because it uses its own stabilization techniques, accounting for them is not enough to completely explain how a style becomes autonomous from the local and microsocial incidents that gave birth to it. Its persistence demands some conditions about people and their place in nature. The style is situated in the midst of people, it responds to the needs, ideological interests or curiosity of some of its members. It begins being propelled by all sorts of social vectors and it is inseparable from the institutions that develop it. It is people who think, reason, investigate. Styles are legitimized by institutions. The cause of an event in the trajectory of a style of scientific reasoning is subject to social analysis.[13] Social history must be evoked not only to explain the origin of each style, but also its continuation, expansion and revitalization. In this sense, styles are contingent, possible thanks to the historical, institutional, economic, etc. conditions. The issue that must be questioned in this respect is: how investigations have been carried out and what was done to establish canons of reasoning and veracity. Let us note that Hacking prefers to speak about reasoning and veracity,[14] instead of reason and truth, because both rationality and veracity have a history, whereas the second pair does not.

[13] In his article "Statistical Language, Statistical Truth and Statistical Reason: The Self-authentification of a Style of Scientific Reasoning" (1992c), for instance, Hacking shows how the development of the statistical style is marked by numerous social events.

[14] As has been noticed, Hacking takes the notion of veracity from Bernard Williams. Williams, in his book *Truth and Truthfulness: and Essay in Genealogy*, claims that the concept of truth is universal. It has no history. It must be implemented by any community whose members speak, create, state. Veracity does have a history. In the history of the West it becomes possible to tell the truth about new kinds of things, in new ways and responding to new standards. The book is a genealogy of veracity.

Styles must be analyzed from the perspective of its cognitive history to be able to account for how the innate capacities universally available were discovered and developed in specific times and places, and later constituted the ways in which investigation about the world and ourselves is carried out. Hacking believes that we not only discover facts and techniques but we also discover, at specific historical moments, ways of thinking, ways of reasoning, which are part of our nature. Each moment in the development of a style is the result of the work of some collective, but even so, there is no collective associated to each style of thinking unless by collective we understand the human race. This is why the study of scientific reason in part of anthropology and not of sociology.

As a style matures, it becomes less modelled by interests and more an unquestionable resource on which to map these interests if objectivity is expected. Each style has its own sources and its own journey, it evolves at its own pace and reaches maturity in its own time. When this happens and it becomes fixed, it no longer needs support or rhetoric to acquire self-confidence and generate its own norm. Stabilization techniques gradually become autonomous from conditions as the style develops.

The statistical style, according to Hacking, for instance, clearly illustrates the interests, rhetorical mechanisms, institutions, modes of legitimation, uses of power, etc., involved in a style. But it also shows how all these elements become increasingly irrelevant as the style becomes more robust, until it culminates as an autonomous mode of objectivity about an extensive class of facts, with its own authority and usable as a neutral tool for various projects.

2.6 Style and Positivity

There are different styles of reasoning and each one of them fixes its own domain the sense of the statements that belong to it. Certain statements can only exist within a certain style of reasoning. Not for nothing does Hacking speak about the style as the space of possibilities for the emergence of certain objects and concepts, and therefore, of statements that deal with them. In his article "Statistical Language, Statistical Truth and Statistical Reason: the Self-authentification of a Style of Scientific Reasoning", when discussing the different stages of the statistical style,

Hacking (2006: 413–422) takes from Williams the following outline, which according to him is perfectly suited to his proposal of styles, insofar as it makes it possible to show how they originate, crystallize and develop historically:

1. A change of conception of that which is to tell the truth about X
2. This significant change is produced in the Y century, and its icon is Z
3. Those who act according to the new style are neither more rational nor better informed than their predecessors. Those who stick to the traditional practice do not have confused ideas nor the opposite convictions with regards to their successors.

he claims that each one of the periods shows, among other things, how certain statements become possible in each specific moment of the development of the style, statements that did not have a clear sense, nor a defined truth value, until the right time (1992c: 142).

There is an interaction between the surrounding social causes and the organization of the reasoning, which becomes vital.

In Foucault, the discursive formation also determines what statements can be uttered. For him, statements are propositions considered from the perspective of their conditions of existence, not from the logical or grammatical perspective. A statement is the modality of existence of a set of signs that makes it possible to refer to objects and subjects and to establish a relation with other formulations.

The statement is the ultimate element of discourse. It is not a structure but

> [...] it is a function of existence that properly belongs to signs and on the basis of which one may then decide, through analysis or intuition, whether or not they 'make sense', according to what rule they follow one another or are juxtaposed, of what they are the sign, and what sort of act is carried out by their formulation [...] a function that cuts across a domain of structures and possible unities, and which reveals them, with concrete contents, in time and space. (Foucault 1969: 115, 1972: 86–87)

It is a function and not a linguistic unit—such as phrases and propositions—in the sense that for a series of signs to be a statement it is necessary for it to be related to other series of signs, thus constituting the field associated with the statement. That is to say, it has no reality as a statement before its inclusion in a rule-governed system. But it is a function also in another sense: a series of signs is a statement because it has a place, a role, within a system. It is a regularity. It is a function that consists in regularizing singularities tracing the curve that runs through the vicinity of said singularities.

Hacking takes a further step and attempts to underscore that each style, besides proposing statements that cannot be uttered before the existence of the style itself, establishes whether they can be candidates to be true or false. Thus, speaking about the probabilistic style, he remarks:

> [...] the official statistics of every nation just did not exist at the beginning of the period under scrutiny, 1821. Not only were the sentences not uttered, but also they could not have been understood. We take for granted that most of the sentences are either true or false. No one will dispute the fact that sentences such as these were not inscribed in 1821. I urge that they did not have truth values. (Hacking 1992c: 143)

Or, as Davidson (2004) remarks apropos this same subject, 150 years ago, the psychiatric theories of sexual identity disorders, the statements about sexual perversion—homosexuality, fetishism, sadism and masochism—were not false but they were not even possible candidates to truth or falsehood. Only with the birth of the corresponding style there emerged the categories of proof, verification, explanation, that allowed these statements to be true or false.

Foucault's archaeologist does not occupy himself with the question of whether statements are true or not. Foucault describes an open logical space where a certain discourse takes place. Hacking is also interested in the space where a certain discourse is possible, but he is also interested—and perhaps more so, and once again, surely due to his analytic training—in the fact that it is the style that establishes if these statements are candidates to truth or falsehood.

The idea of positivity is perhaps one of Hacking's most frequent references when he acknowledges his debt to Foucault. The archaeological history makes it possible to establish what Foucault calls a positivity, a space where it is possible to establish whether Buffon and Linnaeus spoke about 'the same thing', deploying 'the same conceptual field', confronting each other in 'the same battlefield' (1969: 166, 1972: 126).

He attempts to establish the kind of positivity that defines a field where eventually there could unfold formal identities, thematic continuities, translation of concepts, polemic games: to play the role of a historical a priori.

Positivity is the historical-empirical substratum of discourses. It is the set of material conditions that make the existence of discourses possible as specific practices. A discourse always has material conditions of enunciation that go beyond its lexical or logical rules, which imply its mode of existence, enunciability, transmission, appearance and disappearance.

Hacking's style of scientific reasoning generates the possibility that, for instance, the concept of probability be thinkable. In Foucault's words,

> This a priori is what, in a given period, delimits in the totality of experience a field of knowledge, defines the mode of being of the objects that appear in that field, provides man's everyday perception with theoretical powers, and defines the conditions in which he can sustain a discourse about things that is recognized to be true. (Foucault 1966: 171, 2005: 172)

This concept of positivity appears in *Maladie mentale et personalité* (1954), by Foucault, and it constitutes the conceptual foundations on which he would later build his investigation on the birth of the clinic. It is not a transcendental principle, it is not Kant's absolute a priori, whose conditions were universally applicable, necessary restrictions for any possible experience. It is an a priori relativized by history. Its conditions are contingent to the particular historical situation and they change according to the time and the domains of knowledge. It is, at the most, a transcendental function exercised in the very space of discursive practices. It is in relation to this domain that it can be said that positivities constitute the historical a priori of discourses. It is not a condition of validity for some judgements but a condition of reality for some statements, it is the condition for their emergence, their way of being, their transformation, their subsistence, their coexistence with others. It is the condition of an already given history, which is that of things actually said (Foucault 1969: 167, 1972: 127). It is not a becoming, but a being. The a priori is always given as something already constituted, it is given from the beginning in its definitive, finished, totalized form.

In the historical a priori two of the fundamental requirements of archaeology can be found: that of establishing a certain regularity and historicizing it to the maximum possible extent. If there was no regularity, relation or possibility of comparison between statements, it would not be possible to describe discursive formations. For this reason, the positivity of a science must be considered a form of regularity, an a priori that determines whether a statement belongs to it or not.

The a priori determines the mode of existence of discourses in their singularity as events. A reality that is in itself historical given the changes suffered by the elements to which the positivity ascribes its regularity. That is to say, positivity is not just a historical a priori in the sense that it is a condition of possibility of the happening of discourses in history, but is itself subject to the changes and transformations that constitute history. It does not proceed on shallow discursive surfaces but rather in complex depths that determine the conditions of appearance of discourses. Far from being a motionless instance that tyrannizes human thought, it is changing and we ourselves end up changing. But it is unconscious: contemporaries have always ignored where their own limits were, as we ourselves cannot perceive ours.

2.7 Style and Truth

Truth or falsehood and a style grow together. A proposition can be considered true or false only when there exists some style of reasoning and researching that helps determine its truth value. What the proposition means depends on the ways in which its truth can be established. There are no statements that already possess truth conditions waiting for someone to discover how to prove them. It is true that some things are true and others false, that there is a real world, that the truth value of statements is external to the style, it does not depend on how we conceive them. That said, this is completely consistent with saying that its truth conditions are the product of the style of reasoning to whose domain they belong, because there is no truth or falsehood in the matter, regardless of the style of reasoning (Hacking 1992c: 135).

According to Hacking (1982: 49) nothing is true or false, but thinking makes it so, and this is what interests him, that the sense of a proposition *p*, the sense in which it points to truth or falsehood, is based on the appropriate style of reasoning for *p*. The rationality of a style of reasoning, as a path that leads to the truth of a kind of proposition, does not seem to open up to independent criticism insofar as the true sense of what is established by the style depends on the style itself. Is it, then, a vicious circle? No, because reasoning does not mean logic, it does not mean preservation of knowledge, but rather the style leads to the possibility of it being true or false, it creates the possibility, the relevant truth conditions, whereas deduction and induction merely preserve it. They make it possible to jump from truth to truth, but they do not give us the original truth from which to depart, and moreover, they take as a given the kind of statements that have a truth value. The style of reasoning is different because it falls on it to generate new kinds of possibilities.

Despite what we have said, there are some statements that do have truth conditions independently of the style. Regarding them, it will be a theory of truth as correspondence that would be able to offer a true clue of its meaning. Whereas this theory has lately received strong objections, Hacking has no issues with a theory of truth as correspondence whose terms designate concepts on a basic level, which can be called pre-style statements. They are statements that sometimes are inferred from the elements of proof, and that we can tell whether they are true or false just by observing. In contrast, there is a number of typically complex questions which, as I have said earlier, can only be answers by means of reasoning. It makes sense to ask about them only within the acceptably reasoned modes towards their answer. The kinds of statements that acquire positivity in a style of reasoning are not well-described by a traditional theory of truth as correspondence.

Hacking considers that there is no theory of truth nor a semantics that applies to the complete set of empirical statements investigated in science. He rejects any uniform and complete aim. He objects to the first dogma of traditional Anglophone philosophy that a uniform theory of truth or of meaning would apply to a complete language. The truth condition of some statements is determined by the ways in which we reason about them. And a style becomes a standard of objectivity because it has the virtue of producing truth (Hacking 1992c: 135). The truth is that found in such and such way. This is recognized as truth by how it is found out. And how does one know that a method is good? Because it provides the truth.

No conjectures are formulated to later check if they are true. Devices are invented to produce data and isolate or create phenomena and for which it is true a network of theories of different levels. What is more, phenomena can only count as such when the data can be interpreted in the light of the theory. An item of knowledge is true when it adjusts to the data generated and analyzed by instruments and devices modelled by topical hypotheses. And it is the style of reasoning that establishes it. So is developed a curious, made-to-measure adjustment between ideas, devices and observation, a coherence between thoughts, actions, materials and marks. These elements,[15] which intervene in the laboratory sciences, can be modified (one by one or all at once) to lead to some kind of agreement. In so doing, the truth of the world is not read. Usually, there is no pre-existent phenomenon that the experiment informs. Phenomena are constructed. The theories of the laboratory sciences cannot be directly compared to the world (Hacking 1992a: 30), but they persist because they are true with regards to the phenomena produced or created by the devices in the laboratory, and measured by instruments also designed by humans. There is no previously organized correspondence between theory and reality that must be confirmed. Theories are at the most true for the phenomenon obtained by means of instrumentation with the aim of finding a good fit with the theory. They are true for the laboratory phenomenon. The process of modification of the work, the instruments—material or intellectually—provides the glue that keeps the intellectual and the material worlds together. It is what stabilizes science.

Hacking defends his ideas about statements, objects and styles, under the spirit of his mentor, Gottfried Leibniz. Far from implying any kind of relativism, claims Hacking, the doctrine that the styles of reasoning are self-authenticating is part of an explanation of what he calls objectivity. Each way of investigating introduces its own criteria of evidence, proof and demonstration. Each one determines the criteria

---

[15] These elements are grouped in: ideas (questions, base knowledge, systematic theory, topical hypotheses, device modelling), things (targets, sources of modification, detectors, tools, data-generators) and marks (data, data evaluation, data reduction, data analysis and data interpretation). They all have something in common: they are plastic, potentially modifiable resources. Each one of them can be modelled and adapted to adjust it to the others. Hacking (1992a).

for "telling the truth" applied within their own realms. This leads to a radical thesis about veracity and objectivity. A style of scientific reasoning is not relative to anything. It does not respond to some pre-existing criterion of objectivity, nor does it determine a norm of objective truth. It is the norm. And this does not mean that a style cannot be modified or abandoned, it means only that a style cannot be directly refuted for showing that it establishes falsehoods.

Despite these claims by Hacking, it is not clear whether his project implies an epistemic relativism or not. Considerations on these questions are not univocal in secondary literature. Kusch (2010, 2011), for instance, claims that Hacking is not successful in his attempts to distance his proposal from epistemic relativism. I agree with this author in that Hacking's early writings account for a proposal close to relativism, but in his later works he has explicitly remarked that the idea of styles of reasoning does not lead to a relativist position. In spite of this, Kusch affirms that Hacking's proposal does invite epistemic relativism, understood as the vision that at least some facts related to epistemic justification are relative to different epistemic practices, and that, despite the fact that they different kinds of practices may develop, they are, in a certain sense, equally valid. In the same sense, Baghramian (2004) claims that Hacking's proposal is an example of relativism because what counts as evidence is internal to a particular style of reasoning. For his part, Bueno (2012: 658) shows how Hacking would actually have the resources to block this relativism, insofar as his proposal allows for the possibility that the same propositions be evaluated by different styles of reasoning. This is to say, according to Bueno, that at least in some cases there is a common standard between the different styles that would allow the analysis of said propositions. The investigations that use multiple styles and common standards provide the specific context in which a choice could be made, precisely because they invoke those relevant common standards.

If there is an interesting theory of truth to discuss, writes Hacking in 1982, that is the one proposed by Foucault:

> 'Truth' is to be understood as a system of ordered procedures for the production, regulation, distribution, circulation and operation of statements. 'Truth' is linked in a circular relation with systems of power which produce and sustain it, and to effects of power which it induces and which extend it. (1984: 46)

For Foucault (1980: 132), truth is not the set of truths that must be discovered and accepted but the set of rules according to which the true and the false are separated and attach specific power effects to truth. Truth is circularly linked to the systems of power that produce and sustain it, and with the effects of power it induces and that accompany it. There is no supreme instance of truth at all, but rather it is the set of proceedings that make it possible to utter, on each moment and one by one, statements that will be considered true. The universal subject of truth is but an abstract subject. The subject is in fact a qualified subject, constructed in and by means of institutional instances.

The form of objectivation is not the same according to the kind of knowledge dealt with. In spite of the desire to be objective, every change of knowledge carries in itself a change of its object. There is no truth as correspondence, there are only singularities. Humans cannot have access to the whole truth because it does not exist anywhere.

In this context, it will only be considered that they tell the truth, that they are received within the game of the true and the false, those that speak in conformity to the discourse of the moment.

Contrary to what might be thought, the aforementioned position does not trivialize the word 'truth'. Situated on the level of the proposition, on the inside of discourse, the division between true and false is neither arbitrary nor modifiable, nor institutional, nor violent.

Neither is this banal for Hacking, who in spite of not wanting a theory of truth at all, considers that truth is necessary in science. Or, in its place, the true.

2.8 An Innovation with Respect to Crombie: The Idea of Crystallization

Some historians, like Crombie, sustain continuity and do not see in it room for radical breaks. Others, on the contrary, see the mutations and do not see room for continuity. Hacking is eclectic, he understands continuationism in a particular way, which includes mutations and catastrophes, more specifically, breaks he calls crystallizations.

Hacking claims that, even though styles can develop or be abandoned, they are immune to any kind of refutation, and in this sense, they are stable and cumulative. They evolve. However, throughout history it is possible to observe radical changes in knowledge and in styles of reasoning; acute discontinuities in the ways in which knowledge is acquired.

Styles are assimilated, some are inserted to integrate others, styles of reasoning are accumulated that were once fundamental and later take on a less central role, making the extension of human knowledge possible:"[…] *how much of a science, once in place, stays with us, modified but not refuted, reworked but persistent, seldom acknowledged but taken for granted*" (Hacking 1992c: 29; Foucault 1980: 133). New styles are continually developed, providing new ways of reasoning about nature.

There are several points where the influence of Foucault can be pointed out in this feature of Hacking's style. Firstly, with regards to mutations, discontinuities or displacements in the systems of thought, its importance for Foucault can be underscored. The French philosopher strives to distance the past from the present, to alter the comfortable situation that traditional historians enjoy in the relation of the past with the present, as can be read in this fragment where he reprimands those historians obsessed with the affiliation of ideas, which is a variant of the continuity thesis:

> […] to seek in this great accumulation of the already-said the text that resembles 'in advance' a later text, to ransack history in order to rediscover the play of anticipations or echoes, to go right back to the first seeds or to go forward to the last traces, to reveal in a work its fidelity to tradition or its irreducible uniqueness, to raise or lower its stock of originality, to say that the Port-Royal grammarians invented nothing, or to discover that Cuvier had more predecessors than one thought, these are harmless enough amusements for historians who refuse to grow up. (Poster 1984: 75)

It is about, according to Foucault, detecting the incidences of the interruptions of thought (1969: 11, 1972: 4), of emphasizing the central part played by discontinuity in the history of thought, with a threefold role: it constitutes a deliberate operation of the historian, it is the result of her description, and it is a concept that the work does not cease to specify, and no longer that pure and uniform vacuum that separates, in a single movement, two positive figures. The breaks, dispersions, interruptions, accidents, are not a mere accident but one of the fundamental laws that regulate discursive behaviors. The discursive formation is exposed to constant transformations and dispersions as a result of its coexistence with other discourses, with which it establishes a system of interchange, interconnection, crisscrossing, overlapping and ruptures. However, changes on a theoretical level that do not also imply fundamental ontological, epistemic and conceptual transformations do not represent a change in the discursive formation.

Foucault's emphasis on discontinuity does not mean, however, that there is no room for gradual transformations or continuous developments. The discontinuity is not total, since the theories and the practices are never independent of what had happened before. What is kept, what persists, is usually re-enunciated, formulated in the vocabulary of the new discursive formation.

On its part, in the style of reasoning I have already pointed out that there is room both for the idea of continuity and for that of discontinuity. However, the coexistence of both in it is more marked, even more so if should it be necessary to emphasize one of both, it would be the first one. Styles are fundamentally continuous, they are long-term processes interrupted by what Hacking calls crystallizations and which give place to discrete periods, always within a continuity. This idea of crystallization is mentioned by Hacking as his main innovation to Crombie's list of styles of thinking. This idea, besides, albeit inspired in Foucault's idea of mutation, marks an important difference with the French philosopher. It shows that Hacking is interested in the historical origin of things.

Hacking's objective is to rethink the whole structure of scientific reasoning from what he calls a Leibnizian point of view. Crombie's vision of the history of European sciences favors continuity. Hacking's point of view is the opposite, but complementary. He claims that the history of each style of reasoning has at least a marked moment of crystallization, a fixing point about how to continue in the future. This crystallization takes place usually after centuries of rudimentary precursors. Hacking acquired this habit early on, in *The Emergence of Probability*. Although in *Scientific Reason* (2009: 16) he points out that in fact he does not call the emergence of probability in 1660 a mutation or even a revolution, in *The Taming of Chance* he claims that "*emergence is about a radical mutation that took place very quickly*" (Hacking 1990: 9).[16]

[16] In "Was there a Probabilistic Revolution 1800–1930?" (1987: 45–55), Hacking remarks that whereas there is a sense in which there was a probabilistic revolution during the period mentioned in the title, it is not a revolution in the terms of Cohen's or Kuhn's *The Structure of Scientific Revolution*. There is another Kuhnian concept that satisfies Hacking: that which appears in Kuhn (1961). These revolutions show features such as being pre-disciplinary, implying new kinds of

Works like *The Emergence of Probability* and *The Taming of Chance*, exemplars of the styles of reasoning, have close similarities to what Foucault does, for instance, in *Naissance de la Clinique* (1963) (*The Birth of the Clinic).* In this text, Foucault aims to, according to Castro (1995: 204) make explicit de historical a priori of a transformation that cannot be attributed to a mutation of thematic contents nor to a variation of logical forms, and which concerns the relationship between saying and seeing.

> Nothing in this ancient arsenal can designate clearly what took place at that turning point in the eighteenth century, when the calling into question of the old clinical theme 'produced' [...] an essential mutation in medical knowledge [...] a new distribution of the discrete elements of corporal space [...] a reorganization of the elements that make up the pathological phenomenon [...], a definition of the linear series of morbid events [...], a welding of the disease onto the organism [...]. The appearance of the clinic as a historical fact must be identified with the system of these reorganizations [...] it is a reorganization in depth, not only of medical discourse, but of the very possibility of a discourse about disease. (Foucault 2003: xviii–xix)

The changes, mutations, conditions of possibility, definitions of objects that Foucault describes about medicine and clinic make it possible to draw a parallel with what Hacking claims about the style of scientific reasoning in general and with which he claims, in particular, for example, regarding the emergence of probability in *The Emergence of Probability*.

Instead of speaking about mutation, Hacking speaks today about crystallization in the evolution of a style, and he clarifies what he means by this by reference to his example of probability. In his book of 1975, he claims that this concept emerged around 1660. Faced with the criticisms that remarked that there were ideas of probability circulating for centuries before that date, Hacking replies that the question is that in the Renaissance these ideas appear with a new configuration that was not even thinkable before. The emergence of modern probability is an example of crystallization of a style of scientific thinking & doing. That is to say, a catastrophic event in the sense that it changes things forever and in a short period of time. This said, claiming that it changes things forever does not mean that probability, in this case, becomes an unmodifiable solid crystal. It means that it is irreversible, a point of no return. It is a radical change because it introduces new objects and criteria for truth or falsehood of the statements about these objects.

Crystallization is usually associated, in general, with some legendary and pioneering figure such as Pascal in the probabilistic style, Linnaeus in the taxonomic style,[17] Galileo-Husserl in the style of hypothetical modelling, Darwin in the style

institutions and being related to a great social change. That is to say, to go hand in hand with a change in the attitude towards the world.

[17] Let us remember that Hacking (2010, April 27: 14) expresses that he is unsatisfied with his statement that the taxonomic style crystallized in Linnaeus' time, even though in that moment there was a radical change from which there was no return.

of historical derivation of genetic development, Boyle in the laboratory style,[18] even though each one of them is not, at the same time, more than a player in that "way of life" that is the style. Schaffer and Shapin in the book *Leviathan and the Air Pump: Hobbes, Boyle and the Experimental Life* (1985), which Hacking has taken as a reference to order the myth of origin of the laboratory style, make a strong use of Wittgenstein's phrase "forms of life". They see themselves as speaking about how and when a new form of experimental life was born. They teach Boyle's experimental program as a new language game and a new form of life. They claim to make a liberal and informal use of the notions of Wittgenstein's "language game" and "forms of life", since their intention is to consider the scientific method as an integral part of certain patterns of activity. Like for Wittgenstein the expression 'language game' underscores that speaking about language forms part of an activity or a way of life, the authors deal with the controversies about the scientific method as disputes about different patterns for doing things and organize people for practical ends. Thus, according to them, the experimental program—not just Boyle's laboratory in Oxford but the program of experimentation developed in Europe in the seventeenth century—was, in Wittgensteinian terms, a "language game" and a "way of life".

Hacking claims that if he was to use Wittgenstein's words for his own purposes, he would suggest that the introduction of the laboratory style developed a new language game within a new way of life. As the style evolved, other language games began to be used. Likewise, it can be said that the ways of life where the laboratory style is practiced have also changed, as has laboratory research.

Whereas Hacking admits having inherited from Gaston Bachelard[19] a certain enthusiasm for mutations, he was influenced more specifically by the breaks in systems of thought dealt with by Michel Foucault in *The Order of Things*. All the

[18] The hero of Schaffer and Shapin's book, *Leviathan and the Air Pump: Hobbes, Boyle and the Experimental Life* (1985), is a device, the air pump, that creates effects that did not previously exist in nature, in isolation. It is the artifice that inaugurated laboratory science. The authors provide, according to Hacking, an unparalleled outline of the painful birth of a new experimental form. They show the material circumstances where the laboratory style crystallized and why the philosopher Thomas Hobbes opposed it. Hobbes saw exactly what Boyle was doing: he was changing the conception of what was true, by using new instruments to create new phenomena and establish new statements about them. The question was not a dispute about the relative weight of empirical evidence in front of deductive proof. It was a deeper issue and one with greater consequences; it was about what would count as evidence. It was a dispute between two men that had different styles. Hobbes sensed the authority of the laboratory, but he lost the battle. Laboratory science, which not only observes nature but also intervenes in it, finally arrived. The crystallization of the laboratory style took place.

[19] Sciortino (2017) points out some differences between Bachelard's and Hacking's revolutionary attitudes. Whereas for the latter styles are cumulative, for the former, in certain moments of history, there take place radical revisions of the conceptual foundations of scientific knowledge, where the past does not need to be rejected but to be transformed. On the other hand, for Bachelard there are extinct ways of thinking that are non-scientific; this is to say, Bachelard uses modern science as a standard to judge certain practices as non-scientific. Hacking, for his part, is not normative and is rather reluctant to say that certain ways of thinking are non-scientific.

same, he admits that the latter inherited in his turn from the former, from Georges Canguilhem and Louis Althusser (Hacking 1981: 76). To understand scientific reasoning, claims Hacking paraphrasing Foucault, interruptions are more important than the longer histories of their predecessors. Only when a style crystallizes is it possible to understand how to investigate things by using it.

But there also points where Hacking himself expresses his distance from the French philosopher. One of them is related to his Braudelian[20] conception of style, unlike the Foucauldian proposal, which is not so. This means that whereas a style is evolutionary and can even be eternal, Foucault's episteme is born and dies, in two moments of transformation. The episteme is a space of knowledge that appeared all at once and that, should it disappear, it will also all at once. Moreover, as I have mentioned above, styles can be abandoned without being supplanted by others, whereas an episteme is necessarily substituted by another one.

2.9 Practical Incommensurability

Some styles have been displaced (renaissance medicine, some astrological doctrines) (Hacking 1982: 60); however, they can be understood all the same. But that understanding is not exactly a translation but the learning of chains of contextualized reasoning for them to make sense. To understand is to learn how to reason. It is wrong to center on the translation of texts, which is difficult when new ranges of possibility are found that do not make sense for the style of reasoning that flourished or is flourishing in another culture.

Hacking separates his idea of style of scientific reasoning from Kuhn's notion of incommensurability, closely related to the notion of translation rather than with that of reasoning. Nothing in his project leads to a semantic incommensurability. In any case, what is subject to revolution, mutation and oblivion is knowledge, the content of knowledge, and not the way in which it is obtained. Some styles have been completely displaced and their objectives cannot be recognized, and even though their systems cannot be translated, it is possible to learn their chains of reasoning which, by the way, have little or no sense without recreating the thinking of those who realize them. What is learned is not what they took as true, but what they took as true and false. Translation of truth is irrelevant, what matters is the communication of the senses of thought. The unit of analysis of conceptual change is not the meaning of terms and concepts. It is not conceptual change in itself what is interesting, but the changes in the function fulfilled by the concepts in general in certain environments and epochs.

[20] Fernand Braudel's conception that history is long-term, with a slow pace that regulates the time of economy, but also of states, societies and civilizations. He considers that the short-term of our life is but the surface of the present time, the political, economic, cultural, social, etc. events below which there is a history that stagnates, that moves slowly, a structural history that resists time, that lasts and even persists. Braudel (1969)

It is Foucault's hypotheses that help Hacking understand that it is not particular theories that are incommensurable, but that incommensurability lies in the bodies of discourse and the systems of possibility. As the French philosopher remarks,

> […] the important thing here is […] a modification in the rules of formation of statements which are accepted as scientifically true. Thus it is not a change of content (refutation of old errors, recovery of old truth) nor is it a change of theoretical form (renewal of a paradigm, modification of systematic ensembles). It is a question of what governs statements, and the way in which they govern each other so as to constitute a set of propositions which are scientifically acceptable, and hence capable of being verified or falsified by scientific procedures. (Foucault 1980: 112)

In Foucault's particular case, where in spite of the discontinuities in thought there is a continuity of sorts for certain concepts, has as a consequence that when moving from one episteme to the next, for instance, a total change of world does not necessarily need to take place, in such a way that it would imply that comparisons could not be established.

But the fact there is no incommensurability between theories does not preclude, according to Hacking, that this might be the case in the laboratory,

> […] because the instruments providing the measurements for the one are inapt for the other. This is a scientific fact that has nothing to do with 'meaning change' and other semantic notions that have been associated with incommensurability. (Hacking 1992a: 56–57)

Incommensurability in the sense that there is no body of instruments to take common measurements because instruments are peculiar to each stable science. The use of more powerful instruments, for instance, can have as a result the production and conceptualization of new kinds of data that are not adjusted to the level of precision of which the previous or established theory was capable. A new theory is needed, with new kinds of precision. A new theory, incommensurable with the previous one. Now, could both be true? This wouldn't be the case if one supposes that there is a single true theory that corresponds with the world. But it is not. The mutual maturation of the new theory and the experiment does not displace the old mature theory, already established, which remains true with respect to the data available in its domain. The diverse systematic and topical theories that persist on different levels of application are true for different phenomena and domains of data. New kinds of data can be produced, thoughts that result from the use of the instruments that investigate more deeply within the microstructure, and that cannot accommodate to the level of certainty of which the established theory is capable. Therefore, a new theory is needed, with new kinds of precision.

2.10 Essence of the Style: Classification

Classifications are not only the essence of a style of scientific reasoning, but also something necessary to think about them.

Hacking agrees with Pierre Duhem about how our fundamental explanations of phenomena are unstable, they are revised, replaced or overruled; but the classifications of phenomena, on the contrary, become increasingly stable and broad as science grows.

Nietzsche (2001) also claims that classifications resist throughout time, they evolve, they are criticized, modified or overthrown. But whereas Duhem defends growth and stability, Nietzsche sees a stability he detests. This stability, however, can be overthrown by creation, new names, new kinds. With new names, new objects emerge. Not rapidly. Only with use, with layer after layer of use. This does not mean that the essence of a new object is created; what is created is its skin, its surface, with which one interacts and on which one intervenes. Gradually its body is solidified, and finally one has the sense of an essence, an essence that has been constructed into a being.

Unlike Nietzsche, for Hacking it is not enough to name in order to create. Nietzsche is, to Hacking's taste, a philosopher still too stuck on saying, and who pays little attention to doing. Naming takes place as situated, in the particular places and times where it happens, in the relations between speakers and listeners, between external descriptions and internal sensibilities. For a name to start to do its creative work authority is required, it is necessary to use it within institutions.

Objects emerge within the style. Ontology was traditionally thought of as an eternal discipline. Nietzsche speaks about the appearance and disappearance of objects, not of being in general but of particular beings, of beings in time. Following him, and also Foucault, Hacking will speak of historical ontology.

It is becoming clear and it will become more evident from now on, how much Hacking's interest in the analysis of historical and situated conditions of possibility owes to Foucault's thought, and how as one advances in the analysis of his work this interest not only persists but is also revealed as the starting point of and link between his investigations. In spite of this, although we have pointed out how many Foucauldian ideas have a direct incidence in as many particular features of the style of reasoning, in this case it is essential to bear in mind that, in every domain, Hacking takes ideas from Foucault but adapts them to his interests. As he himself has stated on numerous occasions: he takes his ideas but he does not copy his vocabulary. He aims to give these ideas his own imprint.

References

Baghramian, M. (2004). *Relativism*. New York: Routledge.

Braudel, F. (1969). *Écrits sur l'histoire*. Flammarion.

Bueno, O. (2012). Styles of reasoning: A pluralist view. *Studies in History and Philosophy of Science, 43*, 657–665.

Castro, E. (1995). *Pensar a Foucault. Interrogantes filosóficos de La arqueología del saber*. Buenos Aires: Biblos.

Collingwood, R. G. (1940). *An essay on metaphysics*. Oxford: Claredon.

Crombie, A. C. (1988). Designed in the mind: western visions of science, nature and humankind. *History of Science, 26*(71), 1–12.
Crombie, A. C. (1994). *Styles of scientific thinking in the European tradition. The history of argument and explanation especially in the mathematical and biomedical sciences and arts*. London: Duckworth.
Davidson, A. I. (2004). *The emergence of sexuality: Historical epistemology and the formation of concepts*. Cambridge: Harvard University.
Fleck, L. (1935). *Entstehung und Entwicklung einer wissenschaftlichen Tatsache. Einführung in die Lehrevom Denkstil und Denkkollektiv*. Frankfurt am Main: Suhrkamp Verlag.
Foucault, M. (1963). *Naissance de la clinique: une archéologie du regard médical*. Paris: Presses Universitaires de France.
Foucault, M. (1966). *Les mots et les choses. Une archéologie des sciences humaines*. Paris: Gallimard.
Foucault, M. (1969). *L'archéologie du savoir*. Paris: Gallimard.
Foucault, M. (1972). *The archaeology of knowledge and the discourse of language*. New York: Pantheon Books.
Foucault, M. (1979). *Microfísica del poder*. Madrid: La Piqueta.
Foucault, M. (1980). *Power/knowledge: Selected interviews & other writings 1972–1977*. New York: Pantheon Books.
Foucault, M. (1984). What is enlightenment? In P. Rabinow (Ed.), *The Foucault reader* (pp. 32–50). New York: Pantheon Books.
Foucault, M. (2003). *The birth of the clinic. An archaeology of medical perception*. London: Routledge.
Foucault, M. (2005). *The order of things. An archaeology of the human sciences*. London/New York: Routledge.
Hacking, I. (1979). Michel Foucault's immature science. *Noûs, 13*(1), 39–51.
Hacking, I. (1981). The archaeology of Michel Foucault. In I. Hacking (2002) (Ed.), *Historical ontology* (pp. 73–86) London: Harvard University.
Hacking, I. (1982). Language, truth and reason. In M. Hollis & S. Lukes (1982) (Eds.), *Rationality and relativism* (pp. 48–66). Oxford: Blackwell.
Hacking, I. (1984). Five Parables. In I. Hacking (2002). *Historical ontology*. (pp. 27–50). London: Harvard University.
Hacking, I. (1987). Was there a probabilistic revolution 1800–1930? In L. Kruger, L. Daston & M. Heilderberg (1990) (Eds.), *The probabilistic revolution* (Vol. 1, pp. 45–55). Cambridge, MA: MIT.
Hacking, I. (1990). *The taming of chance*. Cambridge: Cambridge University.
Hacking, I. (1992a). The self-vindication of the laboratory sciences. In A. Pickering (Ed.), *Science as practice and culture* (pp. 29–64). Chicago: Chicago University.
Hacking, I. (1992b). "Style" for historians and philosophers. *Studies in History and Philosophy of Science, 23*, 1–20.
Hacking, I. (1992c). Statistical language, statistical truth and statistical reason: The self-authentification of a style of scientific reasoning. In E. M. Mullin (Ed.), *The social dimensions of science* (pp. 130–157). Notre Dame: University of Notre Dame.
Hacking, I. (1995). Imagine radicalimente costruzionaliste del progresso matematico. In A. Paganini (Ed.), *Realismo/antirealismo*. Milano: La Nuova Italia.
Hacking, I. (2002). *Historical ontology*. London: Harvard University.
Hacking, I. (2003). *Styles de raisonnement*. Cours au Collège de France. http://www.ianhacking.com/collegedefrance.html
Hacking, I. (2006). *Véracité et raison*. Cours au Collège de France. http://www.ianhacking.com/collegedefrance.html
Hacking, I. (2009). *Scientific reason*. Taiwan: National Taiwan University.
Hacking, I. (2010, April 21). Lecture I. Methods, objects, and truth". [Unpublished]. México: UNAM.

Hacking, I. (2010, April 26). Lecture II. The Second Group of Styles. [Unpublished]. México: UNAM.
Hacking, I. (2010, April 27). Lecture III-A. Taxonomy. [Unpublished] México: UNAM.
Hacking, I. (2014). *Why is there philosophy of mathematics at all?* Cambridge: Cambridge University.
Kuhn, T. (1961). The function of measurement in modern physical science. *Isis, 52*, 161–193.
Kuhn, T. (1962). *The structure of scientific revolutions*. Chicago: University of Chicago.
Kusch, M. (2010). Hacking's historical epistemology: A critique of styles of reasoning. *Studies in History and Philosophy of Science, 41*, 158–173.
Kusch, M. (2011). Reflexivity, relativism, microhistory: Three desiderata for historical epistemologies. *Erkenntnis, 75*, 483–494.
López Beltrán, C. (1997). Foucault y Hacking: una comparación historiográfica. In A. V. Gómez (Ed.), *Racionalidad y cambio científico* (pp. 123–152). México: Paidós-UNAM.
Nietzsche, F. (2001). *The gay science*. Cambridge: Cambridge University.
Poster, M. (1984). *Foucault, Marxism and history. Mode of production versus mode of information*. Cambridge/Oxford: Polity/Basil Blackwell.
Putnam, H. (1978). *Meaning and the moral sciences*. London: Routledge/Kegan Paul.
Sciortino, L. (2017). On Ian Hacking's notion of styles of reasoning. *Erkenntnis, 82*(2), 243–264.
Shapin, S., & Schaffer, S. (1985). *Leviathan and the air –pump: Hobbes, Boyle, and the experimental life*. Princeton: Princeton University.
Vagelli, M. (2014). Ian Hacking. The philosopher of the present. *Iride, 27*(72), 239–269.

Chapter 3
Probability. Books That Smell of Other Books

My work has been seriously influenced by Foucault (or by successive Foucaults) for many years. Books I have written and books I am writing reek of his effect on me.

Ian Hacking (1990b: 70)

Abstract In this chapter, *Probability. Books that smell of other books*, I deal with the node of Probability by means of an analysis of two of Ian Hacking's most representative texts. Regarding *The Emergence of Probability*, I show that Hacking not only uses the Foucauldian archaeological method, but he practically paraphrases its characteristics when explaining the methodology he uses in this book. I also show that while Michel Foucault analyzed the historical conditions of possibility of knowledge for each one of the historical periods he deals with in *The Order of Things*, Hacking tries to show which are the historical conditions that make the emergence of probability possible. In *The Taming of Chance*, Hacking aims to show that analytic philosophy does not need to be the antithesis of historical sensibility and even though he uses the same archaeological methodology, he goes beyond it to reflect the influence of Foucault's genealogical works. This can be seen in what could be called an incursion into Foucauldian biopolitics. Following Foucault's line called "history of the present", in this text Hacking aims to understand how we think and why we seem to be compelled to think in a certain way.

Keywords Probability · The emergence of probability · The taming of chance · The order of things · Archaeology · Genealogy · Historical meta-epistemology · Ian Hacking · Michel Foucault

M. L. Martínez Rodríguez, *Texture in the Work of Ian Hacking*, Synthese Library 435, https://doi.org/10.1007/978-3-030-64785-8_3

From his earlier works, Hacking showed interest in investigating how we came to live in a universe of possibilities where we think about everything in terms of probabilities. We live in a world of probabilities that did not exist in the seventeenth century. Hacking is interested in how this change in our worldview and in ourselves took place. Both texts analyzed in this chapter, *The Emergence of Probability* and *The Taming of Chance*, attempt to tell this story. Neither aims to explain questions about the foundations of statistical reasoning, as does *Logic of Statistical Inference* or *An Introduction to Probability and Inductive Logic*. Both represent a different type of exercise. In the first place, what Michel Foucault called archaeology. Secondly, they are not only illustrative of examples or particular cases of application of style of scientific thinking & doing, but they also constitute the historical background that inspires the notion of styles and brings to light their features and workings in actual practice. The style generates a priori, but in history, the possibility that, for instance, the concept of probability be thinkable. In Foucault's words, it generates the conditions of possibility for the emergence of a discourse in its materiality of events. On the other hand, it also generates the possibility for certain objects to emerge. *The Taming of Chance* clearly shows, as we will see, how departing from the statistical classifications new objects emerge, new people (Hacking 1990a: 3). Or how Adolphe Quetelet, based on his statistical studies about the measurements of the thoracic diameter of Scottish soldiers distributed according to their average, creates a new kind of object: a population characterized by an average and standardized dispersion.

3.1 *The Emergence of Probability* (1975a)[1]

Hacking has mentioned on several occasions the influence that *The Order of Things* had, especially on his two books of 1975: *Why Does Language Matter to Philosophy?* (1975b) and *The Emergence of Probability*. However, as he himself remarks, in the former there were some signs, albeit not too many, of having read Foucault (Hacking 2005: 3), whereas in the second, this influence is much more important. So much so that, in "*Les mots et les choses,* forty years on" (2005), Hacking claims that his book could be called *The Order of Things, the footnote.*

This, however, does not mean that the only book of Foucault that left its imprint in *The Emergence of Probability* has been *The Order of Things*. At the time he was thinking about probability, Hacking also read Foucault's works on madness and said that he was completely fascinated. This led to an important change in its way of doing philosophy. The idea of Foucault's history of the present made him

[1] Hacking says at the end of the Preface to the French edition of this book: "*J'ai eu de la chance: je n'ai pas imité Michel Foucault en choisissant un titre comme The Birth of the Clinic. Une naissance est sans précurseur. Une émergence est une floraison soudaine après presque rien. Mon livre ne parle pas de la naissance de la probabilité, mais de son émergence*" (2002b: 24).

understand the history of probability in a new and different way. The emergence of probability was a history of the present.

In his article "Five Parables"—under the subtitle "Too Many Words"—Hacking claims that the problem is our exaggerated trust in words as the whole, as the stuff of philosophy. In reference to the linguistic turn, he states that there is a subtle linguistic veil on the eyes of some philosophers that makes them read Kant, for instance, as a philosopher of language. But, to avoid the unkindness of speaking about others, he refers to his own work, mentioning as an example his book *The Emergence of Probability* and his conference of 1973 at the British Academy: "Leibniz and Descartes: Proof and Eternal Truths". About them, he says that

> I had been reading Foucault, but, significantly, I had chiefly been reading *Les mots et les choses* (1970), a work that does not so much emphasize mots at the expense of choses, as make a strong statement about how words impose an order on things. (Hacking 1984: 34)

According to his remarks in *The Taming of Chance* (1990a: 9), the central claim of *The Emergence of Probability* is that many of the philosophical conceptions of probability were formed by the nature of the renaissance ideas which preceded immediately the mutation that took place in that field around 1660.

This reference has to do, on the one hand, with what is clearly shown in the first chapters about how the space of possibilities is structured in a way that allows the concept of probability to emerge. On the other hand, it has to do with statements such as Foucault's in the first pages of *The Order of Things* characterizing the book as a study that strives to rediscover on what basis knowledge and theories became possible (1966: 13, 2005: xxiii).

3.1.1 Archaeology in The Emergence of Probability

In my view, what Hacking does in his book is apply this methodology, the archaeological methodology, striving to explain the emergence of a specific concept: probability. Archaeology is a historical method of description of language on the level of statements or discursive formation. It is on these grounds that Foucault will define what he understands by discourse and by discursive practices. As analysis takes a course towards the study of dispositives and practices rather than epistemes, he will situate discursive practices with the framework of practices in general, which includes non-discursive practices, and the center-stage will no longer be occupied by the being of language but rather by its use and its practice in the context of other practices that are not linguistic in nature. This is to say, the issue of the being of language will be substituted in works published after *The Order of Things* by "what it is done" with language.[2]

[2]Apropos the expression "what is done with language", Foucault accepts positively what Austin calls the perlocutionary effect of language as the means that leads to concerted action.

In the second chapter, Hacking explains what methodology he will use in the text and even though he does not call it "archaeology" it is possible to clearly identify some of the distinctive features of the Foucauldian method. On this matter, he says:

> Is better to expose the crudities of one's model at the start, than to conceal a methodology in banal phrases. I am inviting the reader to imagine, first of all, that *there is a space of possible theories about probability* that has been rather constant from 1660 to the present. Secondly, *this space resulted from a transformation* upon some quite different conceptual structure. Thirdly, some characteristics of that prior structure, themselves quite forgotten, have impressed themselves on our present scheme of thought. Fourth: *perhaps an understanding of our space and its preconditions can liberate us from the cycle of probability theories that has trapped us for so long.* (Hacking 1975a: 16) [my emphasis]

The fragment starts by referring to what I consider a constant in Hacking's work: the analysis of the historical and situated conditions of possibility of the emergence of concepts and objects. In the book we are dealing with, this is visible in the way he analyzes the emergence of the concept of probability. He also refers to the historical conditions of possibility of knowledge in Foucault's terms, a question that, as I have mentioned above, the French philosopher approaches in *The Order of Things.*

In the Preface (1966: 13, 2005: xxiii–xxiv), Foucault presents his work more as an archaeology than a history in the traditional sense of the term. It is a history of the historical conditions of possibility of knowledge, which depend on the bare experience of the order and its ways of beings, and that establish what can be said in a certain epoch. The archaeology aims to analyze this experience of order, the intermediate region between the fundamental codes of a culture and the scientific and philosophical theories that explain why there is an order previous to words, to perceptions and to gestures, a more solid, more archaic, less dubious region, always more "true" than the theories that attempt to give it an explicit form, an exhaustive application or a philosophical grounding (Foucault 1966: 12, 2005: xxiii). The region that fixes, insofar as an experience of order, the historical conditions of possibility of knowledge. According to Foucault, all forms of thought involves implicit rules that materially restrict the range of thought, which do not define the existence of reality nor the canonical use of a vocabulary, but rather the order of things. Discovering these rules makes it possible to see how an apparently arbitrary restriction really makes sense in the scheme defined by them. This analysis of what is beyond the control of individuals that live and think at a certain age is key to understanding the restrictions within which people think. Archaeology does not deal with textual analysis, with specific questions about what particular words mean or how certain statements are logical or rhetorically connected. It is situated before this manifest level of specific linguistic use.

But *The Order of Things* is not the only illustrative text in this sense, given that, just as it looks into the conditions of possibility for the emergence of the object denominated "abstract humanity", *History of Madness* and *The Birth of the Clinic* do the same regarding the objects denominated "madness", understood as mental illness and "clinical illness", respectively. Beyond the different perspectives, which make that in the first case the question is about the constitution of some enunciative forms—the human sciences—in the second it is about an object –madness—and in the third by the constitution of a subject-gaze—the physician, the clinic—all three texts converge in one intention and outline a common project (Morey 1983: 33).

For its part, the space of possibilities Hacking talks about resulted from a fundamental transformation from which the concept of probability emerged. Once again, it is possible to see a reference to the French philosopher; in this case, a correspondence with his idea of discontinuity.[3]

In Foucault's case, the mutation of thought or its discontinuity represent a fundamental category both history and the archaeological method (Castro 1995: 26). It forms part of the workings of discourse. It names the set of breaks, dispersions, interruptions. However, as already pointed out, it is not total, since theories and practices are never independent from what has taken place before.

In *The Order of Things* (1966: 64, 2005: 56), particularly, Foucault remarks that discontinuity—the fact that in a few years a culture might stop thinking as it had done so until then and starts thinking about something else and in a different way—certainly opens up to an erosion of the exterior, to this space that, for thought, is on the other side, but about which it has not stopped thinking from its origin.

The traditional history of ideas underscores the continuity of human thought through the centuries, considering it as a homogeneous manifestation of a single mind or a collective mentality. The undermining of this kind of continuity has been part and parcel of this new way of doing history of thought, which questions the privileged role of the human subject. The emphasis on discontinuity is a strategy in its service. It replaces the subject of becoming with the analysis of transformations in their specificity, it is the transformation of the discontinuous, its movement from obstacle to practice, its interiorization in the discourse of the historian that has allowed it to be no longer the exterior fatality that must be reduced but the operational concept which is used. The discontinuous is no longer the negative in historical reading, but rather the positive element that determines its object and validates its analysis.

However, Foucault has made it clear that archaeology deals with changes that take place from one discursive formation to another, but these changes can take place against a background of significant continuities. In this sense, it cannot be distinguished from the traditional history of ideas because it ignores change or continuity, but because it takes discontinuities as seriously as continuities and does not

---

[3] Foucault points out four consequences of the new disposition of history: the multiplication of ruptures, the new importance of the notion of discontinuity, the impossibility of a global history and the appearance of new methodological problems. Foucault remarks in *The Archaeology of Knowledge and the Discourse of Language*: "*One of the most essential features of the new history is probably this displacement of the discontinuous: its transference from the obstacle to the work itself; its integration into the discourse of the historian, where it no longer plays the role of an external condition that must be reduced, but that of a working concept; and therefore the inversion of signs by which it is no longer the negative of the historical reading (its underside, its failure, the limit of its power), but the positive element that determines its object and validates its analysis*" (1969: 17, 1972: 9).

attempt to reduce to the former a series of gradual changes that contribute to a final illumination.

In *The Emergence of Probability* Hacking, for his part, clearly refers to a radical mutation developed very rapidly from precedent conceptions and which gave place to the emergence of probability, *"The preconditions for probability will consist in something that is not probability but which was, through something like a mutation, transformed into probability*" (1975: 9).

Finally, at the end of the selected passage, Hacking alludes to his idea that concepts have a memory and that some of the philosophical problems about these concepts are the result of their history. In his paper "Five Parables", Hacking claims that many of the philosophical problems emerge from the forgotten history of concepts. These emerge at a certain moment thanks to a different ordering of previous ideas that explode or collapse. The incoherence between the previous and later states of ideas many times trigger these philosophical problems, and failing to understand that the source of the conflict is the absence of coherence between the concept and this previous order of ideas that made it possible causes the problem to persist.

In a later work, "Historical Ontology" (1999b: 8–9), Hacking once more remarks that concepts have a memory and that a correct analysis requires accounting for their trajectory and previous uses, because those who fail to understand the history of their own central organization of ideas are doomed to failing to understand how they are used. In other words, he points to the need of knowing the road our concepts have travelled to be able to apprehend and comprehend them.

At the end of the second chapter of *The Emergence of Probability*, Hacking continues to describe his methodology:

> I do not ask any reader to swallow all this. The story told in what follows is of interest even if the methodology that led to it turns out to be silly [...]
>
> To begin with, the probability to be described is autonomous, with a life of its own. It exists in discourse and not in the minds of speakers. *We are concerned not with the authors but with the sentences they have uttered and left for us to read* [...] We are not concerned with who wrote, but with what was said. This attitude will irritate the proper historian. He wants to know how an idea is communicated from one thinker to another, what new is added, what error deleted. *I am more interested when the same idea crops up everywhere*, on the pens of people who have never heard of each other.
>
> My model has other implications. *I tend to disregard the anticipation*, the man who, with subtle interpretation, can be presented as a precursor of the modern way of thought. In prehistory we are not interested in what is rare but in what is common. Common does not mean familiar –it may be utterly bizarre–. For example, I say, with only very slight reservations, that there was no probability until about 1660. How do I know? I have not read every text. There are many texts that no-one fluent improbability lore has ever read. How can I so confidently talk of the beginning of this family emerges permanently in discourse. (Hacking 1975a: 16–17) [my emphasis]

Hacking enters here the realm of the difference between the archaeological method and the traditional history of ideas. He alludes to Foucault's proposal of an analysis of thought independently of the subject or subjects that produce it.

Foucault's archaeology is significant as an alternative to the traditional approach, which sees the history of thought as constituted by the human subject. Unlike the latter, the former turns from the subject towards the conditions that define the discursive space in which the speaker subjects exist. This introduces a fundamental difference between it and the history of ideas that translates into different attitudes towards tradition and innovation. Archaeology is not interested in the great topics of traditional history of ideas, namely: the genesis, continuity, tradition, totalization, teleology or evolution, the influence, mentality of spirit of the times, etc. Even though Foucault does not exclude the discussion of how philosophers, scientists or other thinkers develop and transmit to their successor key concepts and theories, he considers that such points of view, centered on the subject can lend themselves to distortions. It is necessary to leave behind these ready-made syntheses, these groupings admitted before any examination, these links whose validity is accepted from the beginning, to expel the forms and the forces by means of which people's thoughts and their discourses are linked, to accept that, in the first instance, all there is a population of scattered events.

Hacking claims that his study is not about great men but rather about an autonomous concept (1975a, 56); likewise, Foucault does not look for an expression of individuality or of society, the instance of the subject-creator of the works, but he rather defines the discursive practices that run across it. He is not interested in what men have thought, attempted or wished to say, but rather on what has been written, in the exteriority of discourse (Foucault 1969: 182–183, 1972: 139). The aim of reading is not to discover the author's intention but the deep structure of language itself, to go beyond individual awareness to reveal the unconscious episteme that defines and makes personal knowledge possible. Thus, archaeology does not construct the meaning of utterances in terms of the speaker's intention, but posits that the meaning of an utterance is a function of the role it plays in their complete system.

Here is rooted the Foucauldian marginalization of the subject, which is not the same as denying reality or the importance of individual conscience, but rather claiming that individuals operate in a conceptual environment that determines them and limits the ways in which they may or may not have knowledge. It is an invitation to question the category of subject, its supremacy, its foundational function. It does not mean the total exclusion of the subject from history. It cannot mean this, insofar as Foucault speaks about "our" history and insofar as he has repeatedly affirmed that the true topic of his investigations is the subject (Foucault 2000: 281–282; 290; Dreyfus and Rabinow 1983: 209). But archaeology emphasizes that the state on which history is represented is established independently from our thoughts and actions. His project is to offer an internal analysis of thought, but without assigning the thinker a privileged role, and of writing without a privileged role for the writer.

Likewise, Hacking is not interested in who wrote about probability but on the fact that the idea appears everywhere, in the Foucauldian sense that a discursive formation gives place to a family of ideas even when they come from different realms, but they have a single structure in common. He also denies the importance of the analysis of relation of succession, inheritance, anticipation, etc., between different thoughts and thinkers, echoing the anti-presentist traditions.

It is in this context that Hacking points out: "*We are looking for neither precursors nor anticipations of our ideas*" (Hacking 1975a: 9).

Neither is Foucault's archaeology interested in the anticipation of ideas, but rather in the regularity of utterances. A reflection on the historical in a branch of knowledge cannot be satisfied with following the thread of knowledges throughout time; they are not questions of inheritance and tradition, and it neglects to say what made them possible by uttering what was already known before the and what are the new elements they have contributed (Foucault 1966: 221, 2005: 226).

Archaeology aims to see structures that are common to the texts and practices of an epoch, which lie more deeply than the level of meaning, and which make it possible, as Hacking claims, *"Whether motivated by God, or by gaming, or by commerce, or by the law, the same kind of idea emerged simultaneously in many minds"* (Hacking 1975a: 103).

It is in this sense that Foucault analyzes, in *The Order of Things*, the periods he calls Renaissance, Classical, and Modernity, attempting to show how in each of them the disciplines that are the counterparts of the present-day human sciences can be understood in terms of a single episteme, common to all of them. This episteme defines the conditions of possibility of all knowledge, whether it is manifested in a theory or remains silently invested in a practice (Foucault 1966: 179, 2005: 183), making history as an epiphenomenon both possible and necessary; the history of ideas, but also of practices.

Foucault presents the episteme of Renaissance as ordering the world in terms of relations of similitude, and the signs as constituted by their similarity to that which they signify. Knowledge is knowledge of similitudes. Four main figures articulate it: convenience and its proximities, emulation and its echoes, analogy and its chains, sympathy and its attractions. Knowledge is no more than relating a form of language with another form of language. Language becomes itself a part of the world, a subsystem of similarities. It forms part of the great distribution of similarities and signatures. Words and things share the same nature and communicate through it. Knowledge of language is commentary, effort to refer, in the form of interpretation, the similar to the similar.

If the episteme of the Renaissance was dominated by similarity, that of classical times is by representation. It is possible to define the episteme of this epoch by the articulated system of a genesis, a taxonomy and a mathesis. It will no longer be a matter of similitudes, but rather of identities and differences. The word is no longer the infinite commentary of a primitive text inscribed in things, but an artificial system of arbitrary signs, bound to a general theory of representation. Since the Classical epoch the representing sign has been, at the same time, indication of the represented object and its manifestation. The soundness of the language of the thing inscribed in the world is dissolved in the workings of representation. The art of language is a way of making a sign, of signifying at the same time something and arranging signs around it. An art of naming and, at the same time, of capturing that name, of enclosing it and saving it, of designing it with other names, which are their deferred presence. Foucault studies in detail three particular domains of knowledge in this period: general grammar, natural history and the analysis of wealth.

With the decline of representation and the consequent fragmentation of knowledge, between the end of the eighteenth century and beginning of the nineteenth, discourse stops playing the organizing role it had in classical knowledge and language loses the central role it had enjoyed in the episteme of that epoch. In the nineteenth century, the general space of knowledge is no longer that of identities and differences, but that of analogies and successions. Things fold over themselves, outside ordered representation. Languages emerge with their history, life with its organization and autonomy, work with its own capacity for production, and in the space freed by discourse man emerges. Language is now just an object of knowledge among many, although this does not mean that it does not have a special meaning in modern thought. It remains the medium through which any knowledge must be expressed. Foucault discusses two complementary projects undertaken by modern thought in order to obtain some control over language: formalization and interpretation. Both are not radically opposed: they have a common origin and purpose. Both are rooted in the new status of language as historical reality and as an object of knowledge.

Life, work, language, therefore, are fundamental concepts of knowledges that do not share points of contact among them, but which archaeology aims to relate and articulate, questioning the similarity between these three domains and whether they might have been affected by the same type of transformation. The heterogeneity of discourses vanishes in the presence of a more important homogeneity that reveals the compatibilities and coherences of an epoch, as well as the mutations and incompatibilities between different epochs.

The question of innovation is of no interest to archaeology. Neither is it the search for a coherence underlying the apparent contradictions—the task of the traditional history of ideas—because archaeology rather than explaining and overcoming contradiction aims to describe its discursive conditions of possibility.

At the beginning of *The Order of Things*, Foucault wonders about the possibility of Borges' (1998) classification in "El idioma analítico de John Wilkins", and poses the hypothesis that will guide him along the text: "[...] *in every culture, between the use of what one might call the ordering codes and reflections upon order itself, there is the pure experience of order and of its modes of being.*" (Foucault 1966: 12–13, 2005: xxiii).

It is the specific nature of this bare experience of order what makes Borges' classification impossible.

From this guiding thread, Foucault aims to find that starting from which knowledges and theories have been possible, according to what ordering space constituted the knowledge, against the background of which historical a priori and in which element of positivity ideas could have emerged, sciences could have constituted themselves, rationalities could be formed. He tries to bring to light the episteme within which knowledges roots its positivity and exhibits a history of their conditions of possibility.

If I take as the basis for a joint analysis of *The Order of Things* and *The Emergence of Probability*, Castro's (1995: 91) remarks that Foucault's intention in this text it is (1) to show that the diverse realms of knowledge studies are isomorphic among themselves and depend on the same historical conditions of possibility and (2) proving that neither philology, nor political economy, nor biology, existed before the mutation that takes place in western thought at the end of the eighteenth century, I can go back to what I have already remarked in Hacking's words and claim that in *The Emergence of Probability* he aims to (1) show that there is a space of possible theories on probability and (2) prove that this space resulted from a transformation in some quite different conceptual structure.

3.1.2 The Emergence of Probability

In the same way as Foucault analyzes the historical conditions of possibility of knowledge for each one of these periods, Hacking attempts to show, especially in the first six chapters of the book, which are the historical conditions that make the emergence of probability possible.

> We should not ask, why did people fail to study these objects? We should ask instead, how did these objects of thought come into being?
>
> All the conjectural explanations I have described try to locate something lacking in pre-Pascalian times. No one denies that arithmetic and nascent capitalism were lacking, nor that one or the other may be essential to the development of probability theory, once probability is a possible object of thought. We should, however, try to find out how probability became possible at all. (Hacking 1975a: 9)

In these first chapters, Hacking shows how the concept of probability emerges, beginning from what he calls the prehistory of chance, departing from the first games of chance like the *talus*, predecessor of dice. The bone of the heel of a runner animal or *talus* is shaped in such a way that when it is thrown on a level surface it can only fall in four ways. Despite the age of this pastime, ideas about probability did not emerge early on, and a mathematics of chance was not known until the Renaissance. In this epoch, what is called probability is an attribute of opinion. Probability mainly means the probability of an opinion. Physicians and alchemists were providers of opinion. Neither the medical science nor alchemy had hopes of being demonstrative sciences.

Only around 1660, many of the necessary ingredients merged to conform the space within which probability as we know it today emerged. Around this time a great deal of people, independently, arrived at the basic ideas of this concept. Even though there had been some anticipations such as the ones we mentioned, it would seem the time was not ripe to give birth to a concept of probability, "our" concept of probability. This lack of maturity was the result, among other things, of the lack of a pertinent concept of factual evidence, whose formation is one of the preconditions of probability. By pertinent concept of evidence, we understand the evidence of things or internal evidence, which is distinguished from the evidence of witnesses and authority. This concept of internal evidence is a bequest of the so-called low sciences of the time (alchemy, geology, astrology, medicine) which, as we have seen, worked only with opinion. They could not carry our demonstrations, and therefore they had to resort to some other form of proof. In the medieval epoch, an opinion was probable if it was approved by the old authority, or at least had abundant testimonies. In the Renaissance a new kind of testimonial was accepted: the testimonial of nature which, like any other case of authority, had to be read. Nature, then, could confer factual evidence—in the modern sense of the atomic, isolated, independent fact, that can serve as an indicator, and even as positive proof, of another isolated, independent fact. However, since it was based on natural signs, only sometimes was it possible to trust it. Probability was communicated by means of what is now called regularities and frequencies, similar to laws. In this way, the connection of probability with stable law-like frequencies is a result of the way in which the new concept of internal evidence came to exist. In this way, only when the sign is transformed in evidence is the space of possibilities given so that a concept of dual probability can emerge.

The observation of signs was conceived as the reading of a testimonial. They were more or less trustworthy. A sign made an opinion probable because it was provided by the best testimonial of all. But signs could also be evaluated by the frequency with which the told the truth. Sign is, therefore, the word that, around 1660, encompasses the new internal evidence and the intentional knowledge of frequency, thus completing the space within which the concept of probability could emerge.

From this moment, probability becomes an object of discussion. Huygens' (1657) book on games of chance is the first printed text on probability. Here, the author sets out to justify a method to evaluate plays. He includes two of Fermat's problems and a game problem provided by Pascal, considered the first significant figure in the modern theory of probability. He obtained this title not only for his well-known correspondence with Fermat on games of chance, but also for his conception of the theory of decisions and because he was instrumental in the demolition of probabilism.

Decision theory attempts to decide what to do when what will happen is uncertain. Given an exhaustive list of possible hypotheses about what the world is like, the observations or experimental data pertinent to these hypotheses, together with an inventory of possible decisions and the different utilities produced by the adoption of these decisions in the different possible states of the world, determine the best decision.

Pascal's colleagues in Port-Royal were the first to talk about measuring something that they specifically called probability. In 1662, in *Logic*, the word probability appears to denote something measurable. This book is one of the first works that provide a new and explicit formulation of the distinction between internal and external evidence. A few years later, John Wilkins proposes to distinguish three classes of evidence: demonstration, testimonial and mixed. The latter is related both with the senses and with understanding, it depends on our own observation and on repeated proof of the issues and instances of actions or things called experience.

For his part, G.W. Leibniz wanted to know what kinds of combinations of conditions, none of them complete in itself, would justify a conclusion of conditional right. He set out to measure the degrees of proof and right in the law with a scale of 0–1, subjected to a crude calculation he called probability.

In those same years, in Paris and London, what we now know as probability and statistics begin to be developed. Whereas several years earlier, in London for instance, a weekly count of baptisms and burials was carried out, John Graunt and William Petty seem to be the first to make good use of these demographic statistics. Graunt designed a mortality chart under the supposition of a uniform mortality. Petty attempted to improve it. In Holland, John de Witt carried out the first serious attempt to obtain fair prices for annuities. He imagined that there was a uniform mortality in the first part of life, but that the rate increased after age 54. John Hudde, however, claimed that to calculate annuities it was necessary to use a uniform mortality. Edmond Halley, in England, based on 5 years of accurate records of death age, was able to work out what at the time was the best mortality table and then combine those data to solve joint income problems. This table was considered the standard for 80 years. Among others, it was used by De Moivre, the best probabilist of his time, who shows that Halley's table could be usefully applied. Having used mortality rates as a real case led him to confirm the model of equal possibilities, the model of equally possible cases (equipossibility). In this sense, both Laplace and Leibniz defined probability as the rate of favorable cases over the total number of equally possible cases. Finally, the emergence of probability was completed with the book *Ars conjectandi* by Jacques Bernoulli, which presents the most decisive conceptual innovations in the early history of probability and demonstrates the first limit theorem. Bernoulli's originality lies in seeing what the notion of certainty meant for probabilities. He thinks that there are two kinds of certainty: subjective and objective. As a consequence of his work, the distinction between the random and epistemic concepts of probability became more important.

The Emergence of Probability, following the model of *The Order of Things*,[4] focuses on the prehistory and the history of probability from a purely conceptual, albeit situated, perspective. As Hacking remarks in his article "Five Parables", philosophy deals with problems—even though it is not just or mainly this-, problems that emerge around concepts and conceptual confusion, frequently due to the lack of knowledge or lack of recognition that concepts must be understood in terms of the words with which they are expressed and the conditions within which they are used. Concepts are *situated* words (1984: 35). These words can express different concepts through changes, for instance, through revolutions, ruptures, mutations or

[4]Whereas the aim of archaeology is to show how institutions, economic processes, social relations, can give place to well-defined kinds of discourse, that is to say not how political practices determine the meaning and shape of discourse, but how and in what way they take part in its conditions of emergence, insertion and functioning, this relation between the discursive and non-discursive domains, present in books from the archaeological stage such as *History of Madness*, disappears completely in *The Order of Things*. In this text, the author neglects non-discursive practices; there is nothing there that corresponds to the asylum of *History of Madness*, to the clinical hospital of *The Birth of the Clinic* or the prison of *Discipline and Punish*. It is an analysis of bodies of discourse. The reason for this is that, according to Foucault, discursive domains do not always follow the structures they share with institutional domains and associate practices, but rather they follow structures shared with other epistemic domains.

epistemological breaks such as those that occur in bodies of knowledge. Words and concepts are not identical.

Some of the philosophical problems about concepts are the result of their history, of lack of knowledge about it. Thus, Hacking's emphasis on the need to know the prehistory of concepts such as probability, chance and determinism, on assuming that the conceptual incoherence that creates philosophical perplexity is a historical incoherence between previous conditions that made a concept possible, and the concept made possible by those previous conditions (Hacking 1981b: 184).

Knowing the prehistory of probability—or of any other concept-, does not mean that the problems it presents will be solved or that they will no longer intrigue us. The history of concepts helps to explain philosophical problems but it does not have the effect of solving them. Removing linguistic confusions will not make the problems disappear (Hacking 1984: 37–38).

3.1.3 The Emergence of Probability, *an Example of Historical Meta-Epistemology*

The Emergence of Probability is one of the first examples of what Hacking calls historical meta-epistemology.

Lorraine Daston has referred to Hacking as one of the best practitioners of the contemporary movement called historical epistemology, and she has underscored the influence that the text discussed here had on her choice of such a label, for a certain kind of investigation she understands as dealing with the historical conditions under which the fundamental epistemic categories of science emerge. About this remark, Hacking has claimed that Daston and her colleagues do not do epistemology since they do not propose, defend or reject theories of knowledge, but rather study epistemological concepts as objects that evolve and mutate. He has also argued that he does not see himself and his work under the label of historical epistemology and that he prefers, in any case, to speak about historical meta-epistemology.

Hacking calls historical meta-epistemology a way of doing history and philosophy of, among other things, the sciences (1999c: 53). Some mixtures of history and philosophy can show how possibilities emerge. Historical meta-epistemology, in this sense, falls under the generalized concept of historical ontology Hacking proposes and which I will develop in the next chapter. As an overview, it can be said that historical ontology studies objects in general, not only things but classes, kinds of persons, ideas and institutions that emerge in history on the basis of certain possibilities. It has to do with the possibility of emergence of some objects and concepts.

Within historical ontology, Hacking calls historical meta-epistemology the examination of the most general concepts used in epistemology. He deals with organizing concepts, related to knowledge, belief, opinion, objectivity, impartiality, proof, probability, argument, reason, rationality, evidence, facts, truth (Ibid: 53).

These are the words used for what Quine called semantic ascent and which Hacking calls in some of his works "elevator words" (1999a: 21). They are words used to say something about what we say or think about the world. They are on a third level. They are concepts that resonate with an emphasis such that they are conceived as objects that lack a history. But, according to Hacking, concepts are not immutable and independent ideas that lie within reach and outside time. These words, together with their adjectives, have experienced substantial mutations of sense and value even though they are often thought about as independent objects, without a history, as if their meanings were stable and transparent, eternal.

Historical meta-epistemology is an attempt to understand these concepts. We cannot think about our thought without using them, and, moreover, they seem to satisfy the following criteria:

1. They structure our thought about the world and organize a whole collection of sub-concepts, practices and values. They are categories of thought, even though Hacking prefers to speak about organizing concepts. They are cousins of Kant's understanding (1999c: 58). Even though they are not permanent, but historical, they play a similar role to Kantian pure concepts that allow us to make judgements.
2. They are historical and situated. They have no other constitution but tradition and use (1999c: 56). They do not exist as a non-temporal resource. They change, evolve, suffer mutations, emerge in the light of new practices or radical transformations of earlier ones.
3. They are unavoidable, possible essentials for the very functioning of our society, our laws, our sciences. They are adhered to us, which does not mean that we cannot change them or that they do not change (1999c: 65).

Their present structure is formed in time and is preserved and modified through it; this is why it can only be explained by examining the ways in which they have been coined and used. Concepts have a memory. A correct analysis requires accounting for their previous trajectories. It is in this sense that Hacking claims that they are situated words.

Hacking has repeatedly insisted that his project is not historical but philosophical, in spite of revealing a sensibility towards history that is not peculiar to the majority of philosophers trained, like himself, in the analytic tradition. A result of this sensibility is its support, on the one hand, on Foucault's history of the present, and on the other, his idea of style of reasoning departing from the notion of style of scientific thinking of Crombie. Thus Hacking has historicized, following Foucault, the forms of reasoning and doing, the forms of telling the truth, the norms of evidence, etc. Regarding this, he has claimed that since writing *The Emergence of Probability* thinks that philosophical problems are created when the historical space of possibilities where our thoughts are organized mutates.

Having said this, why does Daston see Hacking—even though he denies it—as forming part of the tradition of historical epistemology? Because independently of the fact that Daston might label her own work as historical epistemology and Hacking his own work as historical meta-epistemology, there is in the work of both authors a fundamental affinity that is related with the aforementioned historicization and with those that in this book are considered underlying interests of Hacking's work.

It is in this framework that *The Emergence of Probability* can be considered an example of historical epistemology, according to Daston, or historical meta-epistemology, according to Hacking. Because it studies the historical and situated conditions of possibility for the emergence of an organizing concept, it is an unavoidable concept. A concept that came to structure our experience of the world in many ways, which shapes the ways in which we know and based on which there opens a space of possibilities where so many other concepts are constituted. It is a concept without which, as Hacking himself remarks, we cannot conceive the world. This is because the book is a third-level investigation, that is to say, the study of an organizing concept that organizes other concepts of a discourse (epistemological, second-level) which in its turn speaks about other discourses (scientific, first-level) that refer to the world. And it does so taking into account the sites where this organizing concept develops and how it changes.

Hacking can prefer to call this kind of analysis historical meta-epistemology instead of historical epistemology. He can claim that, as he understands it, in the style of Gaston Bachelard, historical epistemology is not the right label for his work on probability and that this very expression, which others use, has acquired a life of its own, but has only a tangential connection with its own intellectual life (Vagelli 2014: 264–265). However, I consider that in spite of all this, it cannot be denied that the kind of analysis Hacking makes, not only in *The Emergence of Probability* but in a good part of his work; continuing, as Hacking himself claimed, Kant's project, but unlike him thinking about scientific reason as a historical and collective product, he shares in general terms important features with the analyses of some of the versions of the so-called contemporary historical epistemology, which provides, at least in my view, grounds for considering his project as the basis for the present-day epistemological proposal.

3.2 *The Taming of Chance* (1990a)

3.2.1 Words in Their Sites

This book, which follows up on the previous one insofar as it shows how we came to live in a universe of probabilities, is a clear example of erudition and historical work strongly influenced by the Foucauldian model, not only regarding methodology but also the profusion of data. Hacking has said of Foucault—I believe that in a characterization that is more about himself than about Foucault—that he "[…] *had an original analytical mind with a fascination for facts. He was adept at reorganizing past events in order to rethink the present*" (Hacking 1981a: 73), and that he likes his longer books more than his short interviews, because the books are loaded with data.

In this work, Hacking aims to show that analytical philosophy does not need to be the antithesis of historical sensibility. On the contrary, and taking up the idea developed in the previous section, he considers that the organization of concepts and the difficulties that emerge from them sometimes have to do with their historical origins. But in this text Hacking not only resumes this thought and reinforces his idea that concepts are situated words, but also that, in some sense, the content of the book originates from the application of this philosophical idea to a specific historical field: the erosion of determinism and the emergence of chance.

Hacking, as well as considering that many philosophical problems are essentially historically constituted, sees in them not only a question of grounding but also of analysis and genesis in the manner of what he calls the Lockean imperative: to understand our thoughts and our beliefs by accounting for how they originate.

As I mentioned apropos *The Emergence of Probability*, taking the project of philosophical analysis of concepts seriously requires, according to Hacking, a history of words in their sites, in order to understand not only what the concept is but what it has been. Speaking about situating words alludes to the utterances—proffered or written—where they appear, but it also means situating them more extensively in terms, for instance, of the institution, the authority or the language in which they are expressed. To do the history of a concept is not merely to discover its elements, but mainly to investigate the principles that make it useful or, eventually, problematic. Even though accepting the more specific conjecture that the modes in which the conditions for the emergence and the changes in the use of a word also determine the conditions in which can be used can result in a complex methodology, this is what, as we have seen, Hacking aimed to do in *The Emergence of Probability* with the concept of probability, and what he theorized in the third part of his article "Five Parables". For its part, *The Taming of Chance* sets out to analyze the conditions of possibility of the emergence of the contemporary conceptions of chance, determinism, information and control, how these conceptions were formed and how the conditions of their construction limit our current ways of thinking. This is what Hacking understands by philosophical analysis, and he claims that he knows only one philosophical model supported by this kind of investigation: some works by Michel Foucault.

Along this line and under the model of the so-called Foucauldian history of the present, Hacking aims to understand how we think and why we seem compelled to think in a certain way (Hacking 1990b: 71). He resorts to history to explain and even undermine concepts we use nowadays, and that we consider inevitable. The specific details of the origin and transformation of concepts such as normal, chance, criminal, perverse, demented, are important to understand them and to understand what makes them problematic. Whereas before this analysis it was possible to see that these concepts were problematic, now we can also see why they are so. Each one of them has been determined by a specific history and their analysis leads us to a local historicism of sorts that deals with particular fields and triggers reflection and action.

Many social problems are closely related to philosophical problems. Sociology provides studies about kinds of behavior—social kinds and, in a certain way, moral kinds such as child abuse—which culminate in a philosophical and historicist question about how the shaping conditions of the corresponding concepts determine their logical relations and moral connotations. This has a strict relation to the idea of how the invention of a classification of people affects not only how we think, how we treat them and how we try to control them, but also how they see themselves. It has to do with evaluation, value creation, and in some cases with social problems about a kind of people, which are consequently regarded as subjects of reform, control, isolation, discipline, etc.

In the case of child abuse, for instance, this kind has been molded in its present form and is a example of how an "absolute value", or what *prima facie* is an "absolute error", has been constructed in front of our very eyes. This category –like many other moral categories—requires an analysis and an understating in the aforementioned philosophical and historical sense. It is a kind that mixes fact and value, and for this reason, because it is evaluative, it has different effects on the investigator than those natural classes might have. In this sense it is an intrinsically moral subject. But it is also extrinsically meta-moral, because it can serve to reflect on the evaluation itself. This reflection can be carried out, according to Hacking, only by tracing back the origin of the idea of child abuse.

In this same sense, but in another text, *Rewriting the Soul*, Hacking also shows that sometimes there are sharp mutations in systems of thought, redistributions of ideas that establish what later seems inevitable, unquestionable, necessary. In this book, Hacking uses multiple personality disorder as a microcosm of what was thought and said about memory in the last half of the nineteenth century and of what is thought and said nowadays, with the purpose of investigating, for example, why something—in this case trauma, memory—seems inevitable, why a set of diverse interests underlies memory. In order to attempt an answer to these and other questions, he observes what happens with memory and multiple personalities mainly in France between 1874 and 1886, when the structure of the modern sciences of memory appears and strengthens. The occurrence—during these two decades—of important changes in ideas persuaded Hacking that this period was a radically formative moment for the idea of memory and that it is not an accident that precisely in this lapse the word trauma –previously used only to signify bodily wounds—should have acquired a new meaning, begun to be applied to psychological damage, and became intimately related to multiple personality. The fact that we do not wonder how trauma became a wound to the soul and that it is taken for granted that memory is a key to the soul, shows that what we think about it as inevitable, invisible and a priori.

On the other hand, Hacking in this text will attempt to answer the question—suggested by Foucault's historicized Kantianism—of how this configuration of ideas emerges and how it has made up and shaped our life, our customs, our science (Hacking 1995: 16). In this respect, I understand that it is the right time to clarify why I speak about historical and situated conditions in Hacking's case. In this way, I aim to point out an important difference between Hacking, Foucault, and Kant. Foucault has repeatedly underscored the Kantian influence in his philosophical work, explicitly situating his work in Kant's critical tradition, which implies an analysis of the conditions under which certain relations of subject to object are formed or modified and a demonstration of how these conditions constitute possible knowledge. Nevertheless, in speaking about historical a priori, Foucault meant to point out an important difference between his notion and the Kantian a priori. Whereas the latter refers back to universally applicable conditions of possibility of knowledge, to necessary restrictions to any possible experience, fixed in time, Foucault does not refer to any transcendental instance. The adjective historical in the Foucauldian expression pretends precisely to move away from this search of transcendental conditions of possibility and to stick to its regular, albeit contingent, historical forms. The historical a priori looks to establish the historical conditions of utterances, their conditions of emergence, their specific way of being, etc., insofar as they are part of an already given history, because it is a history of things actually said. They are conditions of actual experience, not of any possible experience. Foucault claims, with Kant, that our thoughts and experiences occur within fixed categorical limits, but he adds that these limits are contingent, a product of our history and changing from one epoch and domain of knowledge to another. "*Where Kant had found the conditions of possible experience in the structure of the human mind, Foucault does it with historical, and hence transient, conditions for possible discourse*" (Hacking 1981a: 79).

Insofar as Kant's search of universal conditions of possibility of knowledge requires an invocation of methods that go beyond empirical studies, Foucault's search does not go beyond that which is available to the empirical methods of historiography. Foucault's empirical origin of thought is opposed to Kant's transcendental origin of knowledge. Along this same line, Foucault has also moved away from Kant in relation to the anthropological problem, alluding directly to the concrete existence, its developments and its historical contents. His conditions are on the side of the object, on the side of historical formation, of a complex network of practices, and not on the side of a universal subject.

Neither could Hacking be satisfied with the Kantian solution insofar as the truth value of statements are fixed by contingent historical circumstances. Propositions are affected by styles of reasoning which are used in a certain historical context. There are statements that cannot be uttered until the historical conditions for their emergence are given. Therefore, the conditions of emergence of the concepts and the objects cannot be found in the structure of the human mind but are rather historical.

In Hacking, this historical character has the particularity of being a history that does not stay within the limits of an epoch but rather exceeds them. It is in this sense that Hacking speaks, for example, of the prehistory of concepts. Hacking also goes beyond the epochal conditions in the sense Foucault confers on them, as shared by all the knowledges of an epoch. The history of concepts and objects in Hacking are particular trajectories, beyond the fact that their emergence might take place in a certain context or style. In this sense, they are situated conditions. A history that does not attend both to regularity in the Foucauldian sense but rather to the specificity of each history (of a concept or an object). Thus Hacking crosses the barrier of the analytic tradition with regards to the use of history, but does not limit himself to a strict continental use.

3.2.2 The Erosion of Determinism and the Emergence of Chance

If *The Emergence of Probability* clearly shows the influence of *The Order of Things*, it follows its archaeological methods in tracing back the emergence of probability—not only in the high sciences such as astronomy and mechanics, but also in the low sciences such as alchemy and medicine, and in realms of common human life—and pays more attention to discontinuities, *The Taming of Chance*, for its part, can be said to resume the line of work of the previous work, and even though it remains under the same methodology, it goes beyond and reflects the influence of other texts of the French philosopher belonging to his genealogical stage. This becomes evident mainly in what we could call an incursion on the Foucauldian bio-politics and that has been suggested, in some sense, by Hacking himself, in the first pages of the text, when he outlines his project:

> My project is philosophical: to grasp the conditions that made possible our present organization of concepts in two domains. One is that of physical indeterminism; the other is that of statistical information developed for purposes of social control [...] I claim that enumeration requires categorization, and that defining new classes of people for the purposes of statistics has consequences for the ways in which we conceive of others and think of our own possibilities and potentialities. (Hacking 1990a: 5–6)

As I remarked, one of the aims of this book is to show how the collection of numbers and the growth of statistical analysis led some nineteenth century philosophers to abandon a mechanistic view of the world and adopt one based on chance, insisting on the influence of statistics on the human sciences. Statistics have contributed to the formation of laws about society and the character of social facts, and have engendered concepts and classifications in the human sciences. They have been seen merely as providers of information, but also, more interestingly, as part of the technology of power of the modern State. This is how Foucault visualizes it, remarking in an interview that what is interesting about eighteenth century statistics is: first, that they begin to be applied in every aspect of the phenomena that affect populations (epidemics, housing conditions, hygiene, etc.), and that such aspects begin to be integrated in a central problem. Second, that new forms of knowledge are applied: the emergence of demography, observations about the extension of epidemics, investigation about milk-maids and the conditions of breast-feeding. Third, that the establishment of power apparatuses enables not only the observation but also the direct intervention and manipulation in all these areas. Where until at that moment there had only been vague and improvised measures of promotion design to alter situations that were barely known, now begins to develop something like a power over life (Foucault 1980: 226).

On this point, his thesis that from the eighteenth century a new form of power emerged; a strategy of development of medicine and the laws that constitute what he calls bio-politics. In *La volonté de savoir* (1976) (*The History of Sexuality. Volume I: An Introduction),* Foucault describes two poles of development in the exercise of power over life. One of them focused on the body as a machine: its disciplining, the optimization of its capabilities, the extortion of its forces, the parallel increase of its usefulness and its docility, its integration into systems of efficient and economical control; the anatomo-politics of the human body (1976: 183, 1978: 139). All these procedures ensure the spatial distribution of individual bodies and the organization, around it, of an entire field of visibility. It is also about the techniques by means of which these bodies remain under supervision and attempt to increase their useful force by means of exercise, training, etc. The techniques of rationalization and the strict economy of a power that must be exerted, in the least costly way, by means of all the technology we could call disciplinary technology of work, which is introduced from the end of the seventeenth century and during the eighteenth century. (Foucault 1997: 159–160, 2003: 242).

The other pole took shape somewhat later; it was focused on the species body, on the body imbued with the mechanics of life and serving as the basis of the biological processes: propagation, births and mortality, health-care levels, life expectancy, longevity, with all the conditions that might make them vary; the bio-politics of population (1976: 183, 1978: 139).

The biopolitical pole does not have as an object the individual body but rather the multiple body, the population. It refers to the quest which, by the end of the eighteenth century, attempted to rationalize problems that population-related phenomena presented to the governments in terms of health, hygiene, nutrition, sexuality, birth rates, longevity, race, insofar as these became political goals. In that moment, the statistical measurement of these phenomena is put into practice with the first demographics. The objective is to govern not just individuals, by means of a certain number of disciplinary procedures, but also all the human beings gathered in populations. This new form of power will deal with collective phenomena such as: (1) the proportion of births and deaths, reproduction and fertility rates of the population, in other words the demographic phenomena; (2) endemic diseases, in terms of their nature, extension, duration and intensity in the population; (3) old age and diseases that leave an individual outside the realm of the market, of individual and collective insurance; (4) relations with the geographical environment. To sum up, it is the set of mechanisms by means of which that which, in the human species, constitutes its fundamental biological features could be part of a politics, a general power strategy.

Bio-politics inserts itself in the complex dispositive already formed by the pair discipline/control and extends it to a reflection about how to govern. In this sense, it will introduce new mechanisms with a series of functions that differ significantly from the corresponding disciplinary mechanisms. In the mechanisms introduced by politics, the interest will be on principle, of course, on previsions, statistical calculations, global measurement. There will also be an attempt not to modify this or that phenomenon, this or that individual insofar as such, but, essentially, to intervene on the level of the determinations of these general phenomena, these phenomena insofar as they are global (Foucault 1997: 162–163, 2003: 246).

It is not hard to see that in this context the statistics of populations and deviation are an integral part of the industrial State. Many of the categories we use to think about people and their activities were created from an attempt to gather data. Census and other similar bureaucratic forms, in creating new kinds of people to be counted, create at the same time ways of structuring social kinds. Great cobwebs of bureaucracy create infinite ways to count and classify people. Birth, death, disease, suicide, fertility, inaugurate the modern era, the era of statistical data. The statistical style has changed the daily world, a world where all things are registered as probabilities: sex, sports, diseases, politics, cosmic collisions, etc. Hacking approaches this topic not so much as a historian but as a philosopher, and in this sense, his interest is focused on indeterminism and the taming of chance. His hypothesis is that the end of determinism did not come suddenly, but rather was the result of a systematic interaction of a great number of events, some famous and others that went unnoticed. Chapter after chapter of *The Taming of Chance* shows how a relentless and firm erosion of metaphysical determinism begins to take place. Hacking analyzes a series of events that underscore what he calls the "statistical style of reasoning":

1. 1640–1693: the emergence of probability,
2. 1693–1756: the doctrine of chances,
3. 1756–1821: the theory of error and moral sciences I,
4. 1821–1844: the avalanche of printed numbers and moral sciences II,
5. 1844–1875: the creation of statistical objects,
6. 1875–1897: the autonomy of statistical law
7. 1897–1933: the era of modelling and fitting.

Each of these dates is marked by a historical event. In 1693 Bernoulli began the work that will culminate in his celebrated theorem of probability. In 1756 the last edition of *Doctrine of Chances* by De Moivre was published and Lambert began his studies about error, while at the same time the rationalist conception of the so-called moral sciences begins, with Condorcet as its greatest adept. The moral sciences aim to study people and their social relations. Condorcet delimits what would later become two different fields: moral science understood as history and moral science understood as probability, statistics, decision theory, cost and benefits analysis, rational choice theory, applied economics, etc.

In 1821 the first statistical publication about Paris appeared and the department of the Seine, in the context of the birth of a new institution: the Statistics Bureau, which gave place to the avalanche of printed numbers between 1820 and 1840. Until that moment there were no official statistics. During and after the Napoleonic era, counting and measurement became what had to be done. The world became, for the first time, numerical. The enthusiasm for statistics was part of an operation of information and control aimed at eliminating deviation. The fascination for the deviation from the norm, and even the use of the normal to signify what generally takes place, started in the decade of 1820.

Some of the statements that are part of these statistics appear as new laws in accordance with style. In France, for instance, it was the laws of bad behavior that seemed to jump from the pages of official statistics. A range of phenomena that until that moment appeared as free choice—crime, suicide, etc.—is now regarded as subject to inexorable laws. The moral laws are the most fascinating ones, but there was also another group of regularities, such as the biological ones, which had been well known since the seventeenth century: birth and death, which were expressed in terms of probability.

In 1844, Quetelet objectified the meaning of a population. The most important representative of regularity in the nineteenth century transformed the average in a real amount. He transformed the theory of measuring unknown physical quantities into the theory of measuring the ideal or abstract properties of a population. He claimed that quantities –insofar as they were a human attribute—in a population had a distribution, like the error curve, around a value as true as the position of a planet. His "average man" does not refer to the human race in general but to the features of a people of a nation as a racial type. This notion of "average man" led both to a new kind of information about populations and to a new conception of the way to control them.

The idea that human attributes, from physical characteristics to moral characteristics, are distributed along a bell-shaped curve was adopted by statisticians and perfected by Francis Galton, who called it "curve of normal distribution". In 1875, on the grounds of his works, the laws of statistics are not used merely to describe but also to explain phenomena. Galton presents the first statistical explanation of a phenomenon on the grounds of his studies on inheritance. He observes how families of talented and retarded parents have unusual children, but he also remarks that the children of distinguished parents tend to be less exceptional than their parents. Unlike Quetelet—who saw in his average man an aesthetic and moral canon—Galton considered that the average or the normal could be considered an expression of human mediocrity, whereas, for instance, in the distribution of the measurements of intelligence, the deviation located on one of the extremes of the curve could be an expression of genius.

Finally, 1897 is the date of publication of *Suicide*, by Durkheim, a masterpiece of statistical sociology of the nineteenth century.

Each one of these periods establishes the different states of the statistical style and as a whole they show that there was virtually no field of human research that was not touched by the events Hacking calls the avalanche of numbers, the erosion of determinism and the taming of chance. This avalanche does not take place because at this particular time people know how to count in a better way, but because the new kinds of data on population can adjust to the research objectives.

However, there is nothing more anonymous than the bureaucracy of statistics, in the same sense as there is nothing more anonymous than power, according to Foucault. Each new classification and each new computation within this classification are designed by a person or a commission with a direct and limited objective. The population is increasingly classified, reordered and administered by principles, each one of which is presented "innocently" by some bureaucrat. Thus, the characters of Hacking's book—Quetelet, Villermé, Farr, Lexis, Galton, Durkheim, etc.—also imposed their personal character. To this day we still live under this influence, since we are still classified according to the systems they developed. However, tactics are progressively taking shape, without anyone actually knowing what they mean. Finally, the result is a complex interplay of supports, different mechanisms of power.

> Let us not, therefore, ask why certain people want to dominate, what they seek, what is their overall strategy. Let us ask, instead, how things work at the level of on-going subjugation, at the level of those continuous and uninterrupted processes which subject our bodies, govern our gestures, dictate our behaviors, etc. In other words, rather than ask ourselves how the sovereign appears to us in his lofty isolation, we should try to discover how it is that subjects are gradually, progressively, really and materially constituted through a multiplicity of organisms, forces, energies, materials, desires, thoughts, etc. We should try to grasp subjection in its material instance as a constitution of subjects. (Hacking 1981a: 82; Foucault 1980: 97)

3.2.3 Genealogy in The Taming of Chance

The Taming of Chance introduces us –albeit indirectly—in a topic Hacking develops in depth in other texts: making up people. Statistical bureaucracy imposes itself not only because it creates rules but also because it determines the classifications within which people must think about themselves and the actions permitted to them.

> There is a sense in which many of the facts presented by the bureaucracies did not even exist ahead of time. Categories had to be invented into which people could conveniently fall in order to be counted. The systematic collection of data about people has affected not only the ways in which we conceive of a society, but also the ways in which we describe our neighbour. It has profoundly transformed what we choose to do, who we try to be, and what we think of ourselves. (Hacking 1990a: 3)

Thus statistics appear as the trigger of this rich notion that Hacking coins for the human sciences. They are triggers in a double sense. Firstly, because Hacking has said that it was works of this kind that inspired this and other notions he proposes for the human sciences. Second, because in actual fact, according to Hacking, people are made up on the ground of classifications that many times appear in this kind of bureaucratic phenomenon.

It is this approach to the question of power and control—more insinuated than analyzed in depth as a mechanism—which leads us to visualize this text as closer to Foucault's genealogical stage.

The genealogic period in Foucault's work refers to his works on the analysis of the exercise of power—and insofar as statistics are an exercise of power, I understand Hacking's book as influenced by this model. Certainly *Surveiller et punir* (1975) (*Discipline and Punish*) is among Foucault's text the one that makes a clearer use of the genealogical method, even though the point of departure of this trajectory was already outlined, 2 years earlier, in *La société punitive* (*The punitive society*), where he claims that after an archaeological analysis the question now was to make a dynastic, genealogical analysis, approaching filiation departing from power relations (2013: 86, 2015: 83–84).

Discipline and Punish is a genealogy of the modern soul and of a new power of judgement; a genealogy of the scientific-judiciary complex where the power to punish finds its support, receives its justifications and its rules, extends its effects and conceals its exorbitant singularity (1975: 27). He sets out to show how the soul is constantly produced around, on the surface of the body by the workings of the power exerted on it.

The History of Sexuality. Volume I is frequently mentioned as a genealogical book, about how individuals have been led to exert on themselves and others a hermeneutic of desire. For their part, the last two volumes of this collection, even though they are sometimes mentioned in the archaeological sense by Foucault himself, are so in a more attenuated way, insofar as they have to do more with ethics than with its mode of historical analysis. In any case, in spite of these remarks, they do not amount to a radical break with his previous work.

Even though he did not write a methodological work on his genealogy, there is a series of principles that guide this analysis. The principle of discontinuity: to treat discourses as discontinuous practices, without supposing that beneath them there is another grand discourse, homogeneous, unlimited and continuous but repressed or censored, to which speech should be restored. Discourses must be treated as discontinuous practices that sometimes intersect, overlap, but also ignore or exclude each other. Genealogy works from the starting point of diversity and dispersion, the chance of the beginnings and accidents: in no case does it pretend to go back in time to restore the continuity of history; on the contrary, it is devoted to the restitution of events in their singularity. The principle of specificity considers that discourses constitute a violence exerted on things as a practice imposed on them. Finally, the principle of exteriority indicates that one must not go from discourse to its inner and hidden core, to the heart of a thought or meaning, but towards its external conditions of possibility, towards what gives place to the random series of those events and fixates its limits. Works such as *History of Madness*, *The Order of Things* and *The Archaeology of Knowledge and the Discourse of Language* showed how inadequate totalizing categories were to approach historical work. In the first place, the shaping of knowledge requires taking into consideration, besides the discursive practices which he accounts for in works such as *The Order of Things*, non-discursive practices, as well as the interweaving of both kinds of practices. The archaeological method attempted to establish the precise space of discursive production within the very framework of discourse, without resorting to any meta-discursive formation. Discourse is analyzed not in terms of who says what but in terms of the conditions under which these utterances have a definite truth value and, therefore, are susceptible of being utterances. These conditions, which reside in the knowledge of an epoch are, according to Hacking, far from the material conditions—understood both as the proper functioning of the vocal cords and the relations of power that permeate thought—of the production of utterances. Hacking remarks:

> Inevitably, *The Order of Things* looks like an idealist book, reminiscent once again of Kant […] This obsession with words was too fragile to stand. Foucault had to return to the material conditions under which the words were spoken. Not wanting to go back to individual speakers or authors, he at least had to consider the interests which spoken and written words would serve. (Hacking 1981a: 79)[5]

[5] In this same article, Hacking underscores the difference between *The Order of Things* and other texts such as *Discipline and Punish*, on this subject.

It is not the case that language no longer matters. Foucault remains interested in understanding how certain utterances are available as true or false in definite locations of history. But these investigations must be contained within an explanation of the possibilities for action of the sources of power. Thus, the question of power extended the field of interest and he realized that it was necessary to introduce an analysis capable of explaining how discursive and non-discursive practices, utterances and institutions, act and interact among themselves. Let us recall that in Foucault's work, as he substitutes the notion of episteme with that of dispositive and finally of practice,[6] discourse analysis becomes increasingly intertwined with non-discursive analysis of practices in general. In *The Order of Things*, unlike his approach in *History of Madness* and the beginning of *The Birth of the Clinic*, discourse is seen as relatively independent from non-discursive structures. But Foucault gradually distanced himself from this vision until he arrived, according to his own remarks, to a better understanding of the relation between discursive and non-discursive structures and the claim that there is no discourse without power. Discourse emerges within, or departing from, an institution, which plays the double role of censoring but also of reassuring. The institution contains discourse, fixing its limits, assimilating it to its order and in this way reassuring its carrier by making them know that their discourse belongs to the order or legality (Díaz 2003: 78). At the same time, it restricts and coerces them, showing them the direction that their discourse can follow and pointing out the risks that lurk beyond those limits.

Secondly, the category of repression becomes inadequate to account for power relations. Power has a positive epistemic role; it not only restricts or eliminates knowledge but it also produces them. Foucault no longer sees power from the perspective of an institutional judiciary apparatus but as a non-discursive practice, and in this way, it brings to light a new network where knowledge commits itself with political institutions that constitute the densest part of those practices. This power, moreover, is anonymous. The old model of repression speaks of an identifiable someone or some part that organizes people's lives. *The History of Sexuality. Volume I* is a polemic against this model and a proposal of anonymous will. The will in question is not the will of anybody in particular.

The shift from archaeology to genealogy might be considered thus an extension of the field of research to include the study of non-discursive practices and to analyze knowledge in terms of power strategies and tactics. In fact, the idea of genealogy suffered changes. In the beginning, it is presented as a descriptive task, whose

[6] Episteme and dispositive are, in general terms, practices. Episteme is the object of archaeological description; the dispositive is of genealogical description. The dispositive is more general than the episteme, which in some sense could be characterized as an exclusively discursive dispositive. The dispositive is the network of relation that can be established between heterogeneous elements: discourses, institutions, architecture, regulations, laws, administrative rules, scientific statements, philosophical, moral, philanthropic propositions, the said and the unsaid. It is a formation that, at a given historical moment, had as its greatest function that of responding to an urgency. In this sense it has a dominant strategic position. It is always inscribed in a power play but it is also linked to one of the junctions of knowledge, which stream from it but, at the same time, condition it. Thus, the domain of practices is extended from the order of knowledge to the order of power.

object is the effective formation of discourses. Later, it is a form of combat against the power effects characteristic of a discourse considered scientific. Therefore, speaking about power will not be something radically different from speaking about knowledge. On the contrary, it is the link by means of which the archaeological dispositive is extended as genealogy: it abandons the epistemological mask and appears as a political discourse (Morey 1983: 227).[7] This means providing an explanation of the changes in the history of discourse which had been merely described by archaeology. It is not, however, a break, insofar as both rely on the same presupposition: to analyze history without referring to the foundational instance of the subject. *The Order of Things* ends by prophesying a new era in which discourse is not about the thinking subject but only about discourse. A good part of genealogy is that: a kind of history that accounts for the constitution of the knowledges, the discourses, the domains of the object, without referring to a subject that might be transcendent in relation to the field of events or who runs, in their empty identity, through history (Foucault 1980: 117). Neither does it suppose a substitution of one by the other insofar as archaeology continues to play a role in Foucault's later works, only now it is not applied exclusively to discursive practices but also to non-discursive ones, and it is combined with a complementary technique of causal analysis. The demonstration of discontinuity and changes of meaning is still an important task. Although in texts such as *Discipline and Punish* genealogy takes priority over archaeology, Foucault presents modern techniques of punishment by imprisonment in terms of the four main categories of archaeological analysis he had identified in *The Archaeology of Knowledge and the Discourse of Language* (objects, concepts, kinds of authority and strategic lines of action), even though they are now applied not only to language but to practices that go beyond mere linguistic expression and produce physical changes in objects. With the genealogical method there opens up a new level of intelligibility of practices, a level that can no longer be captured by theory alone. At the same time, a new method for deciphering the meaning of such practices is introduced. Using this new method, theory is not only subordinated to practice but it also appears as one of the essential components by means of which their organization takes place. That is to say, it is not a question of opposing the abstract unity of theory and the concrete multiplicity of facts. Neither is a question of disqualifying the speculative element opposing it to the rigor of stabilized knowledge. In fact, it is about putting into play local, discontinuous, disqualified, non-legitimized local knowledges against the unitarian theoretical instance that pretends to filter them, to order them in a hierarchy in the name of true knowledge. Genealogy is the union of erudite knowledge and local memories, which allows for the constitution of a historical knowledge of struggle and the use of this knowledge in present-day tactics. In other words, archaeology is the proper method of analysis of local discursivities, and genealogy is the tactic which, departing from these local

[7] Hacking will not make this shift; as he himself has remarked, his work does not have the political ambition Foucault's work had.

discursivities so described, sets in movement knowledges that did not emerge, now liberated from subjection (Foucault 1980: 85).

In this re-dimensioning of the relations between knowledge and power in the thought of Foucault there is a strong influence of the Nietzschean conceptions of relation, on the one hand, between history and the subject, and on the other, between history and power. Nietzsche has considered power as the essential objective of philosophical discourse, and for Foucault power is a way to approach the topic of the subject. There is no other point where Foucault's thought evokes more obviously that of Nietzsche than his remark that there is an intimate relation between knowledge and power. This leads Foucault to posit that changes in thought are not due, as will be discussed, to thought itself but to social forces that control the behavior of the individuals. Power transforms the main archaeological frameworks that underlie knowledge.

It could be said that the genealogical approximation to history is a question of restituting to archaeology its role of description of discursive and non-discursive practices, to exhibit an essential relation between power and knowledge, and to exploit this relation in order to provide a causal explanation of changes in discursive formations and epistemes. Even though archaeology is capable of describing the system that underlies a practice (discursive or otherwise), it does not adapt to the description of the effects of such practices. It is not a causal, diachronic analysis, but rather a structural, synchronic one. Foucault discusses this limitation in the Foreword to the English translation of *The Order of Things*, where he remarks that he has simply described the systems of thought and did not attempt to explain the changes from one system to the other, from one episteme to another. In this book, episteme functioned as the key to all discourses and history was a succession of epistemes. Each epoch had its own episteme, which was the foundation of all utterances. The episteme functions as a totalizing concept that left Foucault defenseless to confront the fact that he could not explain the change that leads from one episteme to another. These changes are discussed in the genealogical stage on the basis of the relation between power and knowledge.

According to Gutting (1989: 270–271), the simultaneous application of archaeology to discursive and non-discursive practices allows Foucault to establish an essential symbiotic relation between knowledge and power. Like other historians, Foucault attributes the changes in non-discursive practices to a wide variety of economic, political, social and ideological causes. But unlike them, he claims that non-discursive practices are modified by a large number of small inter-related changes instead of a single teleological unified framework. The changes in non-discursive practices that constitute the power structure of a society must be understood as due to a complex and diffuse variety of microfactors, to the microphysics of power. The action of these microcauses can eventually lead to new kinds of non-discursive practices and to a revolution in the correlative discursive practices. Moreover, the objects of these diverse and specific causes are the human bodies. The forces that conduct history do not operate so much on thought, social institutions and the environment, but on the individual bodies. The task of genealogy is precisely to show that the body is also directly involved in a political field. Foucauldian genealogy is a material, multiple and corporeal causal historical explanation. As Foucault remarks in *The History of Sexuality. Volume I*,

> [...] the purpose of the present study is in fact to show how deployments of power are directly connected to the body to bodies [...] Hence I do not envisage a "history of mentalities" that would take account of bodies only through the manner in which they have been perceived and given meaning and value; but a "history of bodies" and the manner in which what is most material and most vital in them has been invested. (Foucault 1978: 151–152)

Power relations find in the body an immediate support. They hide it, mark it, discipline it, torture it, force it. One of Foucault's greatest findings is his ability to isolate and conceptualize the way in which the body becomes an essential component for power relations to operate in modern society. In his analysis, Foucault has diagnosed biopower[8] as the specific kind of power/knowledge of our epoch.

Hacking does not make—in *The Taming of Chance* or other texts—a systematic and explicit analysis of power such as Foucault's. However, as we will see in a later chapter, what Foucault says about power and control can perfectly well be adapted—in my view—to the classification and making up people Hacking discusses. In *The Taming of Chance,* Hacking shows how the double process of erosion of the determinism of the natural sciences and the early development of the sciences of chance and chaos was related to processes of power and control—such as the increasing complexity of social control that took place between the Industrial Revolution and the French Revolution—and the alteration in the concepts of normality and abnormality in the complex social phenomena and processes that emerged.

The moral scientists of the nineteenth century, although they witnessed the great amount of possible variations between people, also found that there are regularities in this variation. They were persuaded that there is a Gaussian curve for any particular attribute of humanity; therefore, the average man is referred to in terms of a statistical dispersion.

On this basis, the idea emerges that it is important to "normalize" people. Normality and statistics, according to Hacking, share a common trajectory. The idea of the normal as opposed to the pathological emerges with greater clarity in the medical discourse around 1800,[9] when data on diseases, criminality, suicide, etc.,

---

[8] The concept of biopower is used here in a general sense, that includes two intersecting axes: disciplines or the anatomo-politics of the body of individuals and the bio-politics of the population. There is another, narrower use, synonymous with bio-politics.

[9] Hacking attributes the origin of the present-day sense of the word to the French physician F. Broussais, who introduced it to medical discourse at the beginning of the nineteenth century. Broussais proposed a qualitative identity between the pathological and normal states, being the former a consequence of a difference in degree in the irritation of organs. In this way, by defining the pathological as a deviation from the normal state, he aimed to inscribed pathology (a discipline of predominantly technical character) within the field of study of physiology, which at the time already enjoyed the status of experimental science (Canguilhem 1966). These ideas were enthusiastically taken up by Auguste Comte, who popularized them and extended the pair normal/pathological together with the thesis that posited its continuity (and, therefore, the dependence of the pathological on the normal) to the study of society. In his political writings he turned a term conceived to speak about the body (the normal state of an organ) in one that could be applied to the

start being recorded, tabulated and published, which leads, in 1830, to the aforementioned avalanche of printed numbers. The normal is equated with the average, and then normality and statistics start sharing parallel careers. One of the determining processes for this and for the contemporary sense of the word "normal" was the adaptation, by Quetelet, of the error curve of Laplace and Gauss to the study of the distribution of human attributes in the population. The expression "normal distribution curve", which we saw in Galton, is symptomatic of a union of trajectories: that of the average ideas or statistical regularity with the pair normal/pathological introduced in medicine by Broussais and popularized by Comte.

As a consequence of its structuring role with regards to the immense majority of classifications created by the human sciences, the meta-concept "normal" exerts pressure on the behavior of people and on how they are conceived by themselves and others.

The development of statistics, probability and the avalanche of numbers converge in the gradual shaping of a landscape and of favorable conditions for the establishment of boundaries between the normal and the pathological, and the incidence of the discourses and practices they give place to. The use of statistics is allied to the establishing of the notions of normal and pathological, which start being applied to subjects,

> People are normal if they conform to the central tendency of such laws, while those at the extremes are pathological. Few of us fancy being pathological, so 'most of us' try to make ourselves normal, which in turn affects what is normal. Atoms have no such inclinations. The human sciences display a feedback effect not to be found in physics. (Hacking 1990a: 2)

In this sense, it can be said that normality plays a fundamental role in the phenomenon of making up people on the part of the classifications of the human sciences, and therefore, also in the looping effect, as we will see in the next chapter.

References

Borges, J. L. (1998). El idioma analítico de John Wilkins. In J. L. Borges (1998) (Ed.), *Otras inquisiciones* (pp. 154–161). Madrid: Alianza.

Canguilhem, G. (1966). Le normal et le pathologique. Paris: PUF.

Castro, E. (1995). *Pensar a Foucault. Interrogantes filosóficos de La arqueología del saber.* Buenos Aires: Biblos.

Díaz, E. (2003). *La filosofía de Michel Foucault* (2nd ed.). Buenos Aires: Biblos.

Dreyfus, H. L., & Rabinow, P. (1983). *Michel Foucault: Beyond structuralism and hermeneutics* (2nd ed.). Chicago: University of Chicago.

Foucault, M. (1966). *Les mots et les choses. Une archéologie des sciences humaines.* Paris: Gallimard.

Foucault, M. (1969). *L'archéologie du savoir.* Paris: Gallimard.

normal State, in the sense of body politic. Emile Durkheim's work inherits this sense of "normal", particularly in *Suicide* and *The Rules of the Sociological Method.*

Foucault, M. (1972). *The archaeology of knowledge and the discourse of language*. New York: Pantheon Books.
Foucault, M. (1975). *Surveiller et punir. Naissance de la prison*. Paris: Gallimard.
Foucault, M. (1976). *La volonté de savoir*. Paris: Gallimard.
Foucault, M. (1978). The history of sexuality. In *Volume I : An introduction*. New York: Pantheon Books.
Foucault, M. (1980). *Power/knowledge: Selected interviews & other writings 1972–1977*. New York: Pantheon Books.
Foucault, M. (1997). *Il faut défendre la société. Cours au Collège de France, 1976*. Paris: Seuil-Gallimard.
Foucault, M. (2000). *Ethics, subjectivity and truth. Essential works of Foucault 1954–1984* (Vol. 1). New York: Penguin Books.
Foucault, M. (2003). *Society must be defended. Lectures at the Collège de France, 1975–1976*. New York: Picador.
Foucault, M. (2005). *The order of things. An archaeology of the human sciences*. London/New York: Routledge.
Foucault, M. (2013). *La société punitive. Cours au Collège de France (1972–1973)*. Paris: Seuil-Gallimard.
Foucault, M. (2015). *The punitive society. Lectures at the Collège de France (1972–1978)*. New York: Palgrave Macmillan.
Gutting, G. (1989). *Michel Foucault's archaeology of scientific reason*. Cambridge: Cambridge University.
Hacking, I. (1975a). *The emergence of probability*. Cambridge: Cambridge University.
Hacking, I. (1975b). *Why does language matter to philosophy?* Cambridge: Cambridge University.
Hacking, I. (1981a). The archaeology of Michel Foucault. In I. Hacking (2002a) (Ed.), *Historical ontology* (pp. 73–86). London: Harvard University.
Hacking, I. (1981b). How should we do the history of statistics? In G. Burchell, C. Gordon, & P. Miller (1991) (Eds.), *The Foucault effect. Studies in governmentality* (pp. 181–195). Chicago: Chicago University.
Hacking, I. (1984). Five parables. In I. Hacking (2002a) (Ed.), *Historical ontology* (pp. 27–50). London: Harvard University.
Hacking, I. (1990a). *The taming of chance*. Cambridge: Cambridge University.
Hacking, I. (1990b). Two kinds of new historicism for philosophers. In I. Hacking (2002a) (Ed.), *Historical ontology* (pp. 51–72). London: Harvard University.
Hacking, I. (1995). *Rewriting the soul. Multiple personality and the sciences of memory*. Princeton: Princeton University.
Hacking, I. (1999a). *The social construction of what?* Cambridge: Harvard University.
Hacking, I. (1999b). Historical ontology. In I. Hacking (2002a) (Ed.), *Historical ontology* (pp. 1–26). London: Harvard University.
Hacking, I. (1999c). Historical meta-epistemology. *Wahrheit und Geschichte. Ein Kolloquium zu Ehren des 60. Geburststages von Lorenz Krüger*. Vandenhoeck & Ruprecht in Göttingen, pp. 53–77.
Hacking, I. (2002a). *Historical Ontology*. London: Harvard University.
Hacking, I. (2002b). *L'émergence de la probabilité*. Paris: Seuil.
Hacking, I. (2005). *Les Mots et les Choses*, forty years on (pp. 1–24). For *Humanities Center*, Columbia University.
Morey, M. (1983). *Lectura de Foucault*. Madrid: Taurus.
Vagelli, M. (2014). Ian Hacking. The philosopher of the present. *Iride, 27*(72), 239–269.

Chapter 4
Making Up People. A Project of More than Three Decades

> *Assim, a minha questão está profundamente relacionada com o que uma vez se chamou de natureza humana, exceto por admitir que nossas naturezas são moldadas pelos nossos conceitos. É uma atitude altamente existencialista –nós não nascemos com essências, mas as formamos no mundo social.*
>
> Ian Hacking (Regner 2000: 10)

Abstract In Chapter 4, *Making up people. A project of more than three decades*, I present the notions Ian Hacking uses to work in the human sciences. Hacking defines himself as a dynamic nominalist, insofar as he is interested in the interaction between classification and the classified individuals, and he vindicates Michel Foucault as antecedent of this nominalism, being interested in the essential role of history in the constitution of its objects, people and forms of behavior. Hence his idea of historical ontology, which deals with the ways in which the possibilities of choice and of being emerge from history and from making up people, that is to say, the ways in which a new scientific classification can make a new kind of people emerge. This interaction between classification and the classified individual results in what Hacking calls the looping effect of human kinds. But making up people, besides, takes place within an ecological niche. Given that in the kinds of the human sciences the aforementioned loop effect presents itself, Hacking proposes the existence of different classes of kinds. In order to illustrate this, I recuperate Hacking's work process on this matter, departing from his original question about whether kinds of people are natural kinds.

Keywords Human sciences · Dynamic nominalism · Historical ontology · Making up people · Looping effect · Ecological niches · Natural kinds · Interactive kinds · Ian Hacking · Michel Foucault

M. L. Martínez Rodríguez, *Texture in the Work of Ian Hacking*, Synthese Library 435, https://doi.org/10.1007/978-3-030-64785-8_4

4.1 Are There Natural Kinds?

Hacking's interest in kinds, and in natural kinds in particular, originates as I have remarked from his interest in human kinds. It is by attempting to characterize the latter that Hacking undertakes the analysis of the former.

To approach the question of natural kinds, particularly in Hacking, it is useful to distinguish between two sometimes undifferentiated aspects: 1. Are there natural distinctions and divisions in the world? and given an affirmative answer to this question, 2. Is this a division in natural kinds? This distinction becomes necessary because, as we will see, it is possible to answer affirmatively to the first question without this meaning to also answer affirmatively to the second. That is to say, it is possible to claim that there are natural divisions between the things of the world without needing to claim that these groupings must be understood as classes.

Many of the discussions about this problem are centered on the natural character of natural kinds and they do not ask whether these divisions constitute classes. Often arguments are used in the sense that natural kinds are essential for the inductive component of science, that the success in predictively induce confirmed generalizations makes it possible to infer what is genuine about natural kinds. However, the fact that kinds are used and that they are a central feature of the success of scientific theories and explanations does not mean that they must be essentials to induction. These discussions, at the most, answer the question of their natural character: natural kinds are so because they are based on natural properties on which inductive inferences are claimed. But they answer this question without answering the other one, of whether these divisions are in classes. A satisfactory reply from a metaphysical perspective should contemplate both aspects of the problem.

In his article "Natural Kinds" (1990), devoted to the analysis of W.V. O. Quine's (1969) homonymous work, Hacking claims – following this philosopher, at least in this aspect—that there are "functionally relevant groupings in nature".[1] He claims that we have reasons to think that some groupings are better than others, for instance, for the carrying out of deductive inferences. This means that there are natural divisions between things, that they are, in some sense, grouped in nature. As Umberto Eco remarks in *Kant and the Platypus* (1999), in the continuum of the content that existed before language effects its vivisections, there are certain lines of resistance or possibilities of flow that make it easier, like the veins in wood or marble, to cut in one direction and not in another. These lines of resistance do not force compulsory senses on the being, they do force on it forbidden senses. There are things that cannot be said about it. In the long and exemplary history of negotiation that is the history of the platypus, for instance – remarks Eco—there was a basis for

[1] Hacking says that Quine gave him with this expression his "*favourite five-word characterization of natural kinds*" (Hacking 2007b: 227). Let us recall that in his article Quine introduces the expression "functionally relevant groupings" when dealing with the problem of induction, but he does not indicate specifically what he means by groupings (Quine 1969: 126). Hacking does not make this clarification either.

negotiation and that was that the animal looked like a beaver, like a duck, like a mole, but not like a cat, an elephant or an ostrich. During these eighty years plus of debate (1798–1884),[2] negotiations were always around the resistances and tendency lines of the continuum, and the decision to admit that certain features could not be denied was due to the presence of these resistances.

But the argument that there are certain natural divisions or distinctions[3] does not imply, at least for Hacking, that these divisions constitute natural kinds. On the contrary, whereas to a certain extent things are grouped in nature, it is people who construct kinds: classes are constructed by people, not found in nature (Hacking 1990:130).

In 1991 Hacking stated he was neither in favor nor against natural kinds.[4] Rather he claimed that the category of natural kind is satisfactory so long as it is kept modest. What does it mean to "keep it modest"? It means that such a notion is productive when it offers indications about facts and distinctions about the world, about us, our experience and our history. When it is useful to do things, when we can use it and with this produce a change, an intervention. Kinds are important for the *homo faber*. History accounts for the importance of natural kinds for intervening in and modifying the world. Natural kinds originated and persist in our interests, according to Hacking, because of what we can do with them and for what the things in these kinds can do for us. However, even though the classificatory legitimacy criterion is the instrumental efficacy of classifications to intervene in the natural world and modify it,[5] Hacking disagrees with a radical nominalism that claims that classes respond exclusively to social interests.

In this as well as in other questions there is a strong pragmatic imprint in Hacking's thought. Kinds are there to do things, not to think or to look for similarities. Faced with Quine's question of whether classes are necessary, Hacking's reply is that we may have an innate instinct for some classes, but not for the notion of class. And in any case, it is not necessary to postulate the existence of such a notion

[2] In 1798, the naturalist Dobson sent the British Museum the tanned hide of an animal that the Australian settlers called *watermole* or *duckbilled platypus* and that we know today as platypus. This animal measures approximately 50 centimeters and weighs 2 kilograms, its body is covered in a dark brown fur, it has no neck, it has a beaver-like tail and a duck-like beak, it is of bluish color on top and rose or marble below, it does not have pinnae, its four legs end in five webbed fingers with claws, it remains underwater a long time, the female lays eggs but she breast-feeds its cubs even though no nipples are seen. In 1884 the debate about the classification of this animal ended when W.H. Caldwell send a famous telegram to the University of Sydney affirming: "*Monotremes oviparous, ovum meroblastic*". Monotremes are mammals and oviparous (Eco 1999: 279–289)

[3] Alexander Bird, in an excellent article about kinds, identifies the school that claims that there are natural divisions among things, such that our categorizations can be successful by equating them with them or not, with the metaphysical thesis he calls weak realism (Bird 2010: 2).

[4] (Hacking 1991a: 110) A curious claim, given that, as we have seen, Hacking already had the idea of style of reasoning since 1975, which would make it odd not to be against natural kinds qua natural kinds.

[5] Not only the natural world but also human kinds, according to Hacking, are causal and instrumental. They are not only part of the system of knowledge but also part of the system of government, of our form of organization (Hacking 1995b: 364).

(Hacking 1990: 134). In the same way we do not need a notion of color to begin with, we do not need an innate notion of class or of similitude, according to Hacking: what we need are classes to work, the capacity of generating classes for doing. Learning a science has to do with learning the kinds of things that fall under their realm. They are what matter in the making of science, in its learning. The construction of new kinds can represent significant theoretical scientific advances. The growth of knowledge, says Hacking, far from what Quine thinks, needs new classes.[6] Classes are neither so important nor so threatening at the beginning, but neither are they negligible in the end.

Unlike his 1991 proposal of a modest notion of natural kinds, in 2007d Hacking claims in a more radical manner the inexistence of what Bertrand Russell called "the doctrine of natural kinds". In "A Tradition of Natural Kinds" (1991a) he attempted to show – bearing in mind the existence of different notions of natural kind—that any characterization of this kind responds to particular contexts; that the notion of natural kind is, as he remarked on other occasions about concepts, a "situated" notion. Hacking's observation did not have the aim of eliminating the notion of natural kind but to distance it from traditional and antagonistic realism and nominalism, and to propose his own alternative: his dynamic nominalism. In 2007, on the contrary, claims that there is a large amount of different analyses about the notion of natural kind directed towards projects unrelated to each other, and that there circulate so many radically incompatible theories about natural kinds that the concept has self-destructed. There are certainly some classifications that are more natural than others, but "*there is no such thing as the class of natural kinds*" (Hacking 2007b: 205). The paradigmatic examples of natural kinds that can be found in the literature – water, sulphur, horse, tiger, lemon, multiple sclerosis, yellow—make up a heterogeneous group for which it is not possible to find a well-defined class, not even some class which, even if it is diffuse, could be used to group them. Hacking does not subscribe to the thesis that there is a single and best taxonomy in terms of natural kinds that represents what nature is like and that reflects the web of causal laws. The idea that a complete and exhaustive taxonomic structure does not make sense even as an ideal to strive towards. A summary examination of the diversity of classes can contribute to see that there is an interesting difference between them. In any case, his thesis that there are no natural kinds does not imply that the expression "natural kinds" is useless. In a pluralist context such as the one John Dupré proposes, for instance, it could be possible to claim that there are ways in which the interest in natural phenomena leads to a choice of certain systems of classification as particularly natural. This kind of use could point to a modest role for the expression "natural kind". This does not mean using it as an absolute classification that divides the groupings in some that are natural kinds and others are not so. It would not have any special relation with nature. It would be used for those classifications

[6] Quine, unlike Hacking, considers that we have an innate ability to use the notions of similitude and class, but that as science advances we get rid of them (Quine 1969: 138).

that according to the context appear as natural, as opposed to those that appear as conventional.

To sum up, whereas at the beginning of the 90s Hacking showed that there were different historical notions of natural kind and that this weakened the idea that there might exist a single kind, in 2007 he attempts to show that this always increasing heterogeneity makes it impossible to even sustain the concept of natural kind, because the expression has been emptied of the metaphysical load ascribed to it in its origins and which gave it its primary philosophical statute.

Despite considering that the notion of class is not necessary, Hacking concedes that a discussion about the notion of natural kinds could be beneficial in two important aspects: with regards to the taxonomic question, that is to say the question about whether natural kinds must be taxonomic, and regarding the relation between human and natural kinds. I will deal with this last question in the next pages, but before I will develop an analysis of his notion of dynamic nominalism.

4.2 Dynamic Nominalism

Hacking's reflections about the classification of people are, as he himself says, a form of nominalism. It is important to point out that Hacking understands this doctrine as meaning that there is no classification in nature that is not mental, in other words, that exists independently from our own human system of nomination. He would like to be able to situate his reflections on human kinds within the great tradition of the British nominalism of Ockham, Hobbes, Locke, Mill, Russell and Austin. However, this nominalism is, in his view, static, unlike his own, interested in the interaction of names and the named: a dynamic nominalism.

Hacking distinguishes three forms of nominalism. In the first place, traditional nominalism, according to which categories, kinds and taxonomies are created and established by humans, and that whereas they can be developed and revised, once established they remain basically fixed and do not interact with what is classified. He considers this nominalism as erroneous and mysterious, because it makes an absolute mystery of our interaction and description of the world.

Secondly, revolutionary (historical) nominalism, attributed by Hacking to Thomas Kuhn[7] – among others—who accounted for how at least one important group of categories emerges in the course of scientific revolutions, of the genesis

[7] Hacking develops this topic mainly in his article of 1993 "Working in a New World". It deals with a polemic interpretation of the Kuhnian proposal, not only because Kuhn himself—as he remarks in "Afterwords" (1993)—does not see his thought reflected in it, but because Hacking seems to translate his own realist and materialist vision of the problem rather than interpreting Kuhn. According to this interpretation, in Kuhn's proposal it is possible to distinguish: a world that does not change, the world of material individuals and another that actually does so by changing the paradigm, the world in which the scientist works and which is made up by kinds that organize populations in incommensurably different ways. The world of material individuals seems unnecessary for Kuhn who, in his latest works, rejects the assumption of a material world –see for instance

and transformation of naming systems. There is a construction of new systems of classification according to certain interests in describing the world, interests closely connected with the anomalies on which the community is centered during periods of crisis. The traditional nominalist assumes that our systems of classification cannot be radically altered. Kuhn changed this vision, showing that categories are successively altered and proposing a less mysterious nominalism, describing the historical processes by means of which new categories of objects and ways to distribute them emerge. In any case, this revolutionary nominalism is not, according to Hacking, strict and true, since for a revolution to be recognized as such, anomalies have to emerge first. Kuhn's revolutionary nominalism is an invitation to historicize the change in categories, but the objects of the sciences, even those described by changing systems of categories, do not appear as historically constituted. If one does certain things, certain phenomena appear. Humans do this. But what happens is constricted by the world. There is an empirical constriction that guides construction.

Like Kuhn's revolutionary nominalism, Hacking claims, Foucault's nominalism is also historical. Hacking does not refer to any particular passage to exemplify Foucault's nominalism, but he is clear in indicating what he means by it, when he claims– for example, in 1981– "*Foucault propounds and extreme nominalism: nothing, not even the ways I can describe myself, is either this or that but history made it so*" (Hacking 1981: 83).

Most philosophers depart from the relation of the philosopher, or of people, with Being, with the world, with God. Foucault departs from history, from which he takes samples to detail his discourse and infers from this an empirical anthropology. The kinds that are supposed to exist are not based on nature, but in the ways in which human beings classify according to the peculiarities of its biological composition, their interests, political agreements or power relations. Ontologically, there is nothing other than variations, singularities, events such as speech acts, enunciation acts or writing, particular people, etc. Heuristically, it is more productive to start from the detailing of practices, of what is done and said, and to make the intellectual effort to make discourse explicit in this way, than to start with a general and well-known idea, since one runs the risk of sticking to this idea, without noticing the ultimate and decisive difference that would reduce it to nothing (Veyne 2009: 19).

The investigation departs from facts and not from a philosophical principle. Foucault looks for the particular conditions of reality; discourses and their dispositive. There is nothing other than the empirical, the historical, or at least nothing authorizes us to state the transcendent or only the transcendental exists.

But Hacking sees something fundamentally different between the nominalism attributed to Kuhn and Foucault's nominalism. In the second, where the objects are people and their ways of behavior, history plays an essential role. This third form of

---

Kuhn's (2000). On the other hand, the epistemic nominalism of Hacking's interpretation seems too modest and realist for the Kuhnian need for classes to constitute the world (Lewowicz 2005: 43).

nominalism, proposed by Hacking and which according to him can also be found in Nietzsche and Foucault, is, as well as historical, dynamic.

Hacking (2002a: 2) considers himself a dynamic nominalist interested in how the practices of naming interact with the things that are named, but he thinks that he could equally be called a dialectic realist, concerned with the interactions between what is and the conceptions of it, concerned about the intervention on this reality. Hacking is also attracted by the alternative of speaking of dialectic realism because the classes of individuals created are real, in any plausible sense of the word. They emerge from a dialectic between classification and who is classified. Naming has real effects on people, and the changes in them have real effects in the subsequent classifications. In any case, we are not concerned by an arid and logical nominalism or a dogmatic realism. Above all, this philosophy is dynamic and dialectic. In the case of child abuse, for example, in which attitudes, definitions, laws and practices destined to deter it evolve in front of our eyes, it is possible to study clearly the dynamics in action, not as a closed history. It is a nominalism in practice opening and closing fields of human possibility, forming and molding people as events develop.

Hacking admits that Nietzsche could have been the first dynamic nominalist when he remarks in *The Gay Science:*

> This has caused me the greatest trouble and still does always cause me the greatest trouble: to realize that *what things are called* is unspeakably more important than what they are. The reputation, name, and appearance, the worth, the usual measure and weight of a thing – originally almost always something mistaken and arbitrary, thrown over things like a dress and quite foreign to their nature and even to their skin – has, through the belief in it and its growth from generation to generation, slowly grown onto and into the thing and has become its very body: what started as appearance in the end nearly always becomes essence and *effectively acts* as its essence! What kind of a fool would believe that it is enough to point to this origin and this misty shroud of delusion in order to *destroy* the world that counts as 'real', so called 'reality'! Only as creators can we destroy! – But let us also not forget that in the long run it is enough to create new names and valuations and appearances of truth in order to create new 'things'. (2001: 69–70).

"Making up people"[8] would be an equally special and fundamental case of this phenomenon. However, some points separate it from the Nietzschean conception. Hacking takes this aphorism with an interest that is different from Nietzsche: to speak about the classification of persons and the interaction between classifications and persons themselves. The aphorism is beneficial in relation to the names of the kinds of people, but not as a doctrine about things. On the other hand, Hacking's sense of the real is too strong for this tendency towards Nietzsche's linguistic idealism. For the former, things depend more on what they are than on what they are named. Nietzsche's aphorism could suggest that names work magically by themselves but, according to Hacking, names are but a part of the dynamics that includes in interaction—at least—also the classified individuals, the institutions, knowledge

[8] "[…] *the ways in which a new scientific classification may bring into being a new kind of person, conceived of and experienced as a way to be a person*" (Hacking 2007a: 285).

and the experts. Names do not work by themselves, as mere sounds and meanings. They function in a world of practices, institutions, authorities, connotations, histories, analogies, memories, fantasies.

Foucault could be seen, according to Hacking (2007a: 4) and although he himself never identified as such, as a more recent defender of dynamic nominalism, understanding homosexuality as a specific way of being that exists only since a particular historical and social time.[9] For him, as for Nietzsche, concepts are evolutions. Therefore, in his genealogy, he attempts to leave anthropological universals aside as far as possible, to question them in their historical constitution.

> In regard to human nature or the categories that may be applied to the subject, everything in our knowledge which is suggested to us as being universally valid must be tested and analyzed. Refusing the universal of "madness," "delinquency," or "sexuality" does not imply that what these notions refer to is nothing, or that they are only chimeras invented for the sake of a dubious cause. Something more is involved, however, than the simple observation that their content varies with time and circumstances: It means that one must investigate the conditions that enable people, according to the rules of true and false statements, to recognize a subject as mentally ill or to arrange that a subject recognize the most essential part of himself in the modality of his sexual desire. (Foucault 1998: 461–462; 1984c: 1452–1453).

If concepts are evolutions, realities are so as well. They do not come from an origin, but they were formed by epigenesis, by means of additions and modifications, and not from a preformation. They have constructed with the passage of time.

Following this line, Foucault wishes to show how the subject is constituted by means of a certain number of practices. It is the concretion of singularities.[10] The subject is not a major fold in the Being. Individuals possess a concrete freedom that can only react against its immediate context (Veyne 2009: 119). Foucault has removed the exclusive and instantaneous right to the subject sovereignty. In fact, far from being sovereign, the subject is constituted, a process Foucault named *subjectivation*.

> What I rejected was the idea of starting out with a theory of the subject-as is done, for example, in phenomenology or existentialism-and, on the basis of this theory, asking how a given form of knowledge [*connaissance*] was possible. What I wanted to try to show was how the subject constituted itself, in one specific form or another, as a mad or a healthy subject, as a delinquent or nondelinquent subject, through certain practices that were also

[9] Foucault's proposal is that the idea of homosexual, as a kind of people, came into being only in the nineteenth century in Europe as a product of bodies of thought and specific medical and legal institutions. Earlier, it was not possible for such a kind of people to exist. This is to say, beyond any time, place and civilization, the sexual orientations might emerge as a question of choice conditioned by social agreements, whatever the orientation, practice, choice and behavior, it won't be possible to produce a kind of people such as the homosexual until after the nineteenth century (Hacking 2002b: 336).

[10] Foucault's objective has been the elaboration of a history of the subject, or rather, of what he calls modes of subjectivation. In this sense, it is important to recall that this history changes style, objects and methodology in the measure Foucault goes from the question of the episteme to the dispositive, and finally, to the practices of self. This leads him to a history of the practices where the subject appears not as a foundational instance but as the result of a constitution, and where the modes of subjectivation are precisely the practices of constitution of the subject.

> games of truth, practices of power, and so on. I had to reject a priori theories of the subject in order to analyze the relationships that may exist between the constitution of the subject or different forms of the subject and games of truth, practices of power, and so on, (Foucault 2000: 290; 1984b: 1537).

The subject is modeled in each epoch by the dispositive and the discourses of the moment, by the reactions of their individual liberty and their eventual aestheticizations. The subject is a result of their time; one cannot become any kind of subject at any time. And it is not enough to say that the subject is constituted in a symbolic system. It is not only in the play of symbols that they are constituted but in real practices, historically analyzable.

In ontological terms, the Foucauldian being is reduced to the succession of discursive practices of knowledge, the dispositives of power and the forms of subjectivation, to a set of discontinuous procedures whose foundation cannot but be indeterminacy (Veyne 2009: 60). All phenomena are singular, every historical or sociological fact is a singularity.

Foucault does not believe in the existence of general, transhistorical truths, since human facts, acts or words, do not proceed from nature, from a reason that would be their origin, nor do they reflect faithfully the object to which they allude (Veyne 2009: 20). There are no natural objects. There are natural substrata that social practices turn into objects. Objects are products of practices, not because of a belief that thought or perception construct reality, but because the given is said, is seen and, to a certain extent, is produced through practices. These, moreover, transform and give place to reality. Practices produce objectivities, and doing archaeology is precisely attempting to discover the practices that support the objectivized.

Hacking, for his part, characterizes dynamic nominalism as a nominalism in action, oriented to new or changing classifications. He deals with the different ways in which classifications interact with the individuals they are applied to. It is the only nominalism that can illustrate how the category and the categorized mutually adjust, and the only one with implications for the history and philosophy of the human sciences, since certain kinds of human beings and actions emerge with the invention of the categories that label them. It is the only one where, as was already remarked, history plays an essential role in the constitution of the objects. Our domains of possibility and our own selves are, in some sense, made up by the names and what is related to them.

Hacking was attracted by this kind of nominalism from the departure point of theories about the homosexual and the heterosexual as kinds of people and by his observations about official statistics.[11] Dynamic nominalism states that there is not

[11] What drew Hacking's attention was how obsessive the moral analysis became around 1820. Rates of suicide, prostitution, vagrancy, etc., generated in their turn subdivisions and regroupings. National and provincial polls showed that the categories used to classify people changed every ten years. This is due partially to the fact that social change creates new categories of people. But the calculation was not a mere report, it was part of an elaborate system, frequently philanthropic, that created new ways "of being" for people and some of them, according to Hacking, were spontaneously adjusted to those categories.

one kind of people that begins to be increasingly recognized by bureaucrats or scholars of human nature, but rather that a kind of people emerges at the same time that the kind itself is invented. As we will see, cause, classification and intervention form a single piece.

In "Making Up People", Hacking exemplifies by means of four categories: horses, planets, gloves and multiple personality. Traditional nominalism is unintelligible for categories such as horse and planet. How could they obey our minds? Gloves are different; they are fabricated. It is not known what came first, if the thought or the mitten, but they evolved together. Hacking's claim about "making up people" is that in some aspects multiple personalities is more like gloves than like horses.

Now, how can dynamic nominalism affect the concept of the individual person? An answer has to do with possibilities, because what one is also include past and future, what could have been done and what will be possible to do. "Making up people" changes the space of possibilities for personality. The creation of a new classification, or the modification of criteria to apply an earlier one, can have effects on the classified individuals that assume or reject the attributes that characterize the new kind. In other words: from the starting point of the creation of new kinds, new possibilities of choosing or existing emerge.

Even though Foucault's poles (anatomo- and bio-politics) or Hacking's vectors – which I will explain later—can be useful to analyze this phenomenon of "making up people", it is necessary to point out that there is no universal theory that can be applied to it, and that such metaphors are mere suggestions that should not make us think that there are two identical histories in this construction. Insofar as dynamic nominalism invites us to examine the complexities of daily and institutional life, it has few chances of being a general philosophy. As a counterpart, even though it does not offer a structure, a system of a general philosophical theory, it can be more interesting for the human question than the arid scholastic nominalism, because it has the merit of stimulating the analysis of the intricacies of real life. There are a plethora of things around us that we not only classify but that we seem to be made to classify them in the way they do. However, it is a project of Hacking's – as it was of Foucault's—to understand how objects are constituted in discourse, how anybody of thought defines only some kinds of objects that fall under certain laws and certain kinds.

On this point it is legitimate to ask ourselves: how does Hacking work this nominalism in relation to the realism he defends, for instance in *Representing and Intervening?* Beyond the fact that Hacking has made it clear that his interest in realism has not been more than a strategy to be able to speak about what really interested him: experimentation, a visible difference can be found between those works where he approaches and explicitly defends a realism of entities and those where he refers to the nominalist question. Hacking deals with the topic of realism mainly in his works on the natural sciences. In these works he appeals to considerations of experimental science to incite realist, anti-idealist and anti-nominalist conclusions. However, it is necessary to remember that this is not due to a belief in the incompatibility of the realist and nominalist doctrines, insofar as he considers that nominalism is about the metaphysical impossibility of classifications, and in this sense the nominalist has no need to deny that there are real things that exist independently of the mind (Hacking 1983: 108). On the other hand, in his works about the human sciences – both in *Mad Travelers* and in *Rewriting the Soul*— he makes it quite clear that he is not interested in answering the question about the reality of multiple personality or of mental disorders in general; as a counterpart, it is in them that the topic of nominalism gains more strength.

Likewise, the problem of kinds does not seem to worry so much when he works with the natural sciences as when he deals with the human ones. It is evident that he imperiously needs kinds to explain the looping effect of human kinds, but not when he is dealing with natural phenomena. Hacking proposes, in this way, different schemas for both kinds of sciences, which agrees with what he expresses in articles such as "The Disunities of the Sciences" (1996b), where he claims that both kinds of disciplines differ and that those differences have to do not with what but with how we know.

But Hacking uses a second strategy that has to do with proposing a different nominalism from the traditional one—for a long time presented as opposed to realism. In *The Social Construction of What?* he says: "*I expect that I am a nominalist because I was born that way. But can I really go whole-hog with Thomas Hobbes and Nelson Goodman? No*" (Hacking 1999a: 233, footnote 23).

Hacking situates himself in an intermediate point between the idea that scientific categories are structurally inherent in the world and that they are only inherent in our forms of representation. Kinds are constructed by people, but there are functionally relevant groupings in nature; things are in some sense grouped in it. If the kinds in which we classify objects had in common only their names, how is it that they fit so well the classified thing? This would be easy to explain for man-made objects, but not for natural phenomena. For this and other reasons mentioned in the text, Hacking needs to propose another notion of nominalism that would satisfy his interests. He cannot think about individuals without kinds. He can only conceive them from the starting point of kinds which, once they touch them, they intervene them and allow us to appropriate them. Meanings are not in language nor in the world in itself, but in a world penetrated by norms and practices. These practices would be present in an evident way in his reflections about interactive kinds and which, for this very reason, seem to prevent him from adopting for them an explanation in terms of natural kinds. Even less so in the terms of an essentialist version of these, since the matter with the kinds of the human sciences is that their members and the properties of these members are in constant flux.

4.3 Historical Ontology

In 1999, Hacking proposed a notion of "historical ontology" that would accompany that other one of dynamic nominalism. In his article "Historical Ontology" he claims to have proposed a new ontology[12] concerned with "[…] *the ways in which the possibilities for choice, and for being, arise in history*" (1999b: 23). In his concept, the notion of historical ontology contributes to thinking as integrating the same family such diverse phenomena as the emergence of probability, the molding of child abuse or the shock produced by transient mental illnesses. This ontology deals with objects, classifications, ideas, people, kinds of people and institutions that emerge in history from certain possibilities; of "[…] *objects or their effects which do not exist in any recognizable form until they are objects of scientific study*" (Ibid:11).

Influenced by Foucauldian thought, this notion is related to the three axes the French philosopher refers to: knowledge, power and ethics. Let us recall that Foucauldian historical ontology has three axes: the axis of knowledge, the axis of power, the axis of ethics. In other words, the historical ontology of ourselves has to answer an open series of questions: how we have constituted ourselves as objects of our knowledge; how we have constituted ourselves as subjects that exercise or suffer power relations; how we have constituted ourselves as moral subjects of our actions (Foucault 1984a: 48–49).

Within the framework of a re-reading of his earlier trajectory from the problem of the subject, and giving it a retrospective sense, in "What is Enlightenment?" (1984a)[13] Foucault renames his task: historical ontology of ourselves. Historical, because it does not establish universal conditions. In Foucault everything is subject to variation. The being is delimited by the two shapes acquired by the visible and the sayable at each moment, by the force relations that undergo variable singularities in each epoch, by the process of subjectivation. Foucault's three great questions: what do I know? What can I do? What am I? could not have a universal answer because they vary with each social formation. And Foucault will praise Kant, in this text, for having been perhaps one of the first philosophers in posing the question "what am I now, a man of such and such time", instead of the questions "I think therefore I am",

---

[12] According to Hacking, ontology concerns two kinds of beings: on the one hand, Aristotelian universals—trauma or child development, for instance—and on the other, the particulars that fall under these universals—this psychic suffering or that childhood development. Universals are not eternal but historical, and their instances, children or the victims of trauma, are shaped and change as universals emerge.

[13] Foucault repeats the distinction between a universal philosophy and the critical analysis of the world in which we live, by means of which he locates its work in the midst of contemporary philosophy, opposing on the one hand philosophy understood as analytics of truth, and on the other, the ontology of the present. Philosophy could consist then, in speaking actuality, in characterizing it negatively, in diagnosing the present, saying what is the present, saying how our present is different and absolutely distinct from anything that is not it. Philosophy becomes then a permanent critique of our historical being.

under its universal form. And even though the "universal" aspect of philosophy does not disappear, the task of the philosopher as a critical analyst of our world here and now is increasingly important.

Historical ontology refers then to three domains of work: an ontology of ourselves in our relations with truth – which allows us to constitute ourselves in subjects of knowledge–, a historical ontology of ourselves in our relations within the field of power – the way in which we constitute ourselves as an active subject that acts on others—and a historical ontology of ourselves in our relations with morality—the way in which we constitute ourselves as an ethical subject that acts upon itself.[14] Under these three domains the I does not name a universal, but rather a set of singular positions. We constitute ourselves in a time and place, using materials whose organization is distinctively and historically shaped.

The presence of the three aforementioned axes restricts, according to Hacking, the possibility that everything that emerges in history could belong to the domain of historical ontology. This is what happens, for instance, with the creation of natural phenomena proposed in *Representing and Intervening* since, even though they emerge in history, they are not historically constituted. The opposite happens with the phenomena studies by the human sciences, submitted to what Hacking calls the *looping effect*[15] of human kinds, that is to say, the interaction between the classification and the classified.

In our attempts to know about people and their behavior – sometimes just for the sake of understanding them, but generally to help them, cure them, socialize them, control them—we constantly create new classifications. These classifications and our knowledge interact with the classified persons, who frequently change or modify their behavior in the light on being classified and known. At the same time, this creation of new kinds implied the availability of new descriptions of behaviors.

The case of the transient mental illness known as multiple personality disorder clearly shows, according to Hacking, the interaction between the experts – in this

[14] Each one of these domains corresponds to one of the periods usually identified in Foucault's work: archaeology, genealogy and ethics. The axis of truth is studied in *History of Madness, The Birth of the Clinic* and *The Order of Things*; the axis of power in *Discipline and Punish* and the axis of ethics in *History of Sexuality.*

However, in all these cases the three axes are present in some measure: "*[...] in these three areas-madness, delinquency, and sexuality- I emphasized a particular aspect each time: the establishment of certain objectivity, the development of a politics and a government of the self, and the elaboration of an ethics and a practice in regard to oneself. But each time I also tried to point out the place occupied here by the other two components necessary for constituting a field of experience. It is basically a matter of different examples in which the three fundamental elements of any experience are implicated: a game of truth, relations of power, and forms of relation to oneself and to others*" (Foucault 2000: 116–117).

[15] "[...] *the way in which a classification may interact with the people classified*" (Hacking 2007a: 286). In "The Making and Molding of Child Abuse" (1991b, but written in 1986), there already appears a mention to this process when Hacking – without calling it "looping effect"- refers to the feedback effect that happens when "*new kinds of people come into being that don't fit the wisdom just acquired, less because the recent knowledge was wrong than because of a feedback effect*" (Hacking 1991b: 254).

case the doctors, who create the possibility of a conceptual space for illness—and the classified, the patients, affected by the classification in their thought, treatment and control, and who act and are, in a way not independent from the available descriptions, thus creating the need to revise the classification and its very criteria of application. Towards 1840, says Hacking, some cases of multiple personality disorder were recorded, but the image that physicians had of this disorder was very different from the present one. Their vision was different because so were their patients, and they were so because the experts had different expectations. Patients of different times tend to adjust themselves to the sense in which they are classified. At the same time, the way in which they do so acquires its own characteristics, which leads to a constant transformation and revision of the classification.

In this process Foucault's aforementioned three axes play a part. Knowledge, because individuals recognize themselves as having a certain kind of behavior and a sense of themselves in relation to the illness. Moreover, new classifications can modify not only the present and the future, but also reinterpret each one's past. If a description was not available in the past, it was not possible to act intentionally under this description. But when it is created, the past can be reorganized in the light of the new categorization. Not only is there a change of opinion about what was done, but also when the understanding and the sensibility change, the past becomes full of intentional actions which, in some sense, were not there when it happened.

The second axis, power, refers mainly to the anonymous conceptual power about the illness, which acts on the life of the patient and of others, because people's interaction with the classification takes place in a matrix of institutions and practices that surround the classification.

Finally, ethics, insofar as events related with illness have to do with values that give place to choices, ways of being and ways of seeing oneself and others. Classifications are related to the creation of values, to evaluation.

It can be observed, then, that Hacking follows Foucault's path in thinking the constitution of subjects not in universalizable terms, but rather as a process that takes place in a time and place, in specific local and historical shapes, and using materials organized in a historically determined way.

In the case of Foucault, why was the historical problem linked mainly to the question about the present? Because of the concept of practice. It is this concept, practice, the only possible continuity of History to the present. In his "What is Enlightenment?", Foucault goes back to the Nietzschean difference between origin and invention, where beginning means the latter, in the sense of human production at a certain point in history. The Foucauldian ontology of the present does not pretend to be grounded in a metaphysics but rather in history (Díaz 2003: 88). In a history of singular and unique events, which even though they might not be subsumed in a teleological chain, they do not respond to simple chance; they are inserted at a crisscrossing of events.

Unlike Kant, for Foucault – as for Hacking—

> [...] the task is not that of fixing an ontologically primitive, definitively 'real' stratum of historical reality, but in tracing the mobile systems of relationships and syntheses which provide the conditions of possibility for the formation of certain orders and levels of objects

and of forms of knowledge of such objects: the uncovering of what Foucault terms a 'historical a priori' (Foucault 1980: 236).

His method does not lead to an ontological search for determination in the last instance; neither does he attempt to infer these diverse orders of events from the causal principle of sufficient reason, but rather to analyze a multiplicity of conditions of possibility, political, social, institutional, technical and theoretical, and reconstruct a heterogeneous system of relations and effects whose contingent interlocking constructs what Foucault calls the historical a priori. In this way, what he achieves is a form of historical intelligibility whose concretion and materiality lies in the true irreducibility of the different orders of events whose relationships he traces (Foucault 1980:243).

This is to say, "[…] *that the historical ontology of ourselves must turn away from all projects that claim to be global or radical*" (Foucault 1984a: 46). It must not be considered as a theory, or as a doctrine, not even as a permanent, cumulative body of knowledge; it should be conceived as an attitude, an ethos, a philosophical life where the critique of what we are is at the same time the historical analysis of the limits that are imposed on us and a proof of their possible transgression (Ibid: 50).

Foucault is interested in showing that what is had not always been, it could not be and it is but the product of several chances and a precarious history. His problem has always been that of the relations between subject and truth: how the subject enters in a certain play of truth.

I mean, how does the subject fit into a certain game of truth? The first problem I examined was why madness was problematized, starting at a certain time and following certain processes, as an illness falling under a certain model of medicine. (Foucault 2000: 289–290; 1984b: 1536–1537).

The closing sentence of *The Order of Things* means precisely that it is possible to say what man is made of, but not to interrogate the being of man or his interiority. Throughout their history, men have never failed to constitute themselves, that is to say, to continuously displace their subjectivity, to constitute in an infinite and multiple series of different subjectivities that would not finally confront us with something that might be *man*.

Like Nietzsche, who attacks the idea of a fixed human nature or essence, Foucault studies the features of human beings that individuals generally think about as fixed, but that a historical study shows to be malleable.

In the same sense, Hacking's ontology does not deal with being in general terms. This is a constant in his work. Let us recall that the experimental argument on which he based realism in *Representing and Intervening* is not an affirmation about the reality of theoretical entities in general, but about the reality of this or that particular entity. What kind of ontology is then Hacking's? One that, as he himself expresses, deals with the particular trajectories of being rather than with great abstractions, that leaves aside the global theoretical debate to focus on some entities in a particular way. It refers more to the space of historical and situated possibilities that surround a person for the formation of their character and that creates the potentiality for individual experience, than to the formation of character itself. It cannot be otherwise, insofar as he underscores once and again that the making up of people does not take place in a general manner but in particular and specific processes, and he claims – in a vision he defines as almost existentialist[16] – that there is no completely fixed human nature about which to discuss.

4.4 Contingentism

This historical ontology, on the other hand, forms part of a wider-ranging contingentist vision, which begins to take shape in Hacking's works on the human sciences of the 80s, but which in *The Social Construction of What?,* for instance, goes beyond the limits of ontology. In "Making up People" Hacking claims not only that when new forms of description emerge, new possibilities for choosing and being appear, but that the possibilities of man to understand himself also change. It is in this sense that he prefers to think about the past as undetermined, since it can be reinterpreted, reorganized and repopulated in the light of new meanings. It is also in this sense that people are not only what they are but what they could have been. In *The Social Construction of What?* this idea of contingency is the first of three conflictive points that he proposes with regards to whether the central elements of contemporary science are contingent or inevitable. The key question in both cases is how to understand conceptual and practical possibilities.

The common denominator is the reiterated notion of possibility, which functions on the grounds of an adaptation of the notions of form and content. As I have already pointed out, Hacking characterizes historical ontology as the ways in which the possibilities for choosing and being emerge in history, and in relation to the space of possibilities that surrounds the individual for the formation of his character. He also uses the notion of possibility to account for famous projects in the history and philosophy of science, such as Ludwick Fleck's – who, according to him, attempts to research what can be thought and how a particular *denkstil* makes certain concepts possible—or Michel Foucault's. In his book of 1999, for its part, his objective is to investigate the ways in which the priority assigned to the military in scientific research from WWII may have affected the shape of present-day science. Hacking is not a contingentist about the content of science, but he is so regarding its shape.

---

[16] In an interview Hacking reminisces: "*I started reading Sartre's Being and Nothingness by the time I was working in the Oil industry, during the summer, and I took pencil notes on every page [...] soon after my arrival in Cambridge my interest for existentialism disappeared. I had start doing other things. We were supposed to be familiar with everything that was conceived as analytic [...] However, there is a certainly a sense in which you could say that my early existentialism is still there, in my philosophy. You can find it for example in my stuff against Quine's idea that there is a scheme of concepts that constrain our ontological possibilities. I rather think, along more Sartrean lines, that there is a scheme of possibilities that shape the set of possible ways of being that is available at a time in history for an individual. All this must be tinged by some vague memories*" (Vagelli 2014: 239–240).

It is also in this sense that laboratory sciences are contingent. Neither the world with which we interact nor past knowledge determine which discoveries of inventions will be made and what will count as discovery. It is contingent which instruments are invented and which phenomena are established or purified, since a science can develop many possible ways. Theories and laboratory equipment are developed in such a way that they interact among each other and they are mutually self-justifying. Such symbiosis is contingent to nature, scientific organization and people.

4.5 Making up People and Looping Effect

The statistics Hacking takes as his point of departure are not, as I have said, a mere report but they create new kinds of people, and in consequence new ways of being and behaving. Given a label, there exists a concrete kind of person to be labelled or, in other words, there is a kind of person which ends up being reified (Hacking 1999a: 27). This phenomenon, characteristic of kinds of people, to which I have already referred as "making up people"[17] owes much to the idea of "constitution of subjects" Foucault refers to in *The Order of Things* when he states that: "[…] *we should try to discover how it is that subjects are gradually, progressively, really and materially constituted through a multiplicity of organisms, forces, energies, materials, desires, thoughts, etc.*" (Foucault 1980: 97; Hacking 1986: 104).

But let us see concretely by means of an example what Hacking means by this:

A. There were no multiple personalities in 1955; there were many in 1985.
B. In 1955 multiple personality was not a way of being a person, people did not experienced themselves in this way, they did not interact in this way with their friends, their family, etc.; but in 1985 it became a way of experiencing oneself, of living in society (Hacking 2007a:299).

When Hacking speaks about making up people he has in mind B. He thinks that there are ways in which many features of people's character and kinds of action are made up. This idea of "making up people" has extended and a good example is, according to Hacking (1986:103), the title of the book *The Making of the Modern Homosexual.* The contributors in this book accept that the homosexual and the heterosexual, as kinds of persons, emerge only towards the end of the 19th century.

Hacking quotes a part of the text, explicit in this respect:

[17] Already in 1991, Hacking says he prefers to use the expression "making up people" rather than "social construction" (1991b: 287). In *The Social Construction of What?* he claims that the discourse of social construction displeases him because *"it is like a miasma, a curling mist within which hover will-o-the wisps luring us to destruction*" (1999a: 101). Moreover, because *"it suggests a one-way street: society (or some fragment of it) constructs disorder (and that is a bad thing, because the disorder does not really exist as described, or would not really exist unless so described)*" (Ibid: 116). By introducing the idea of interactive kind he wants to clarify the existence of a two-way path.

> One difficulty in transcending the theme of gender inversion as the basis of the specialized homosexual identity was the rather late historical development of more precise conceptions of components of sexual identity. It is not suggested that these components are 'real' entities, which awaited scientific 'discovery'. However once the distinctions were made, new realities effectively came into being (Hacking 1986: 103).

If the dynamic nominalist is right about his thesis on sexuality, before the end of the nineteenth century people only had the possibility of being members of the heterosexual kind, because there were no other sexual kinds of which to form part. Analyzing the difference between people and things, Hacking claims that what camels, mountains and microbes are does not depend on words. What happens to the tuberculosis bacillus depends on whether it is treated with the BCG vaccine, but it does not depend on how it is described. It is true that it is treated with a certain vaccine partly because it is described in a certain sense, but it is the vaccine that kills it, not the words. The possibilities of microbes are delimited by nature, not by discourse. The interesting point about human action is that what is done depends on the possibilities of description. Hence if new modes of description appear, new possibilities for action appear as a consequence. The possible ways of being for people emerge from time to time and they appear and disappear. Human actions are actions under descriptions. The courses of action that people choose and their ways of being depend on available descriptions. If the description does not exist, intentional actions under it cannot exist either. The making up of people refers back to the space of possibilities for personality, to which our domains of possibility and us ourselves as human beings are in some sense constructed by names and what is related to them.

But descriptions not only affect what the individual is; they also affect their projects, the kind of person they expect to be, their present, past and future.

Although it has already pointed out that there is no general history about making up of people, at least two common vectors can be mentioned: (1) the vector of labelling, by a community of experts, which creates a reality that some people make their own; and (2) the vector of autonomous behavior of the labelled person, which pushes to create a reality that the expert must face. The histories of the kinds of people are different from those of natural kinds because the former are shaped and molded by interacting with and altering the individuals and the kinds of behavior to which they are applied.

Hacking is interested in knowing how this idea of making up people affects the true idea of what an individual is. Frequently, the creation of a kind provides the space for certain being to adjust to it and, in a certain way, for them to be, thus enabling making up people. It is not the case that first the concept is formed and then the regularities or laws of objects that fall into this category are discovered. The process is interactive. Items are grouped because it is believed that a classification aids comprehension, explanation, judgment and prediction of features of the classified items. Interactive is a concept that is applied to classifications, to kinds, to the kinds that can influence that which is classified.[18] Postulating a classification and its

[18] In later works, Hacking rejects the possibility of speaking about interactive kinds, although he accepts that of interaction.

application to people produces effects on the individuals so classified, who react to the description made of them, modifying their behavior and producing a change in the existing classification to adapt it to the new features of its members. The new classification and theorization induce changes in the self-conception and behavior of the people classified. These changes demand revisions of the classification and the theories, of causal connections and expectations. This process of feedback, or looping effect of human kinds, makes the phenomena studied by the human sciences unstable, mobile objectives, unlike natural phenomena.

Knowledge and classification are intimately related here. The systematic collection of data about people affects not only the ways in which a society is conceived, but also the ways in which peers are described. The invention of a classification for people – and its application— produces numerous effects: it affects how individuals, so classified, are conceived, treated and controlled. It affects how they see themselves. It is strongly related to evaluation, to the creation of values. They are value-laden kinds, kinds of things to be done or not. Kinds that people might want or not to belong to, since human kinds have an intrinsic moral value.

But interaction is not, according to Hacking (2007a: 288–289), merely between the names and the named, but at least between the following elements:

1. The classification and its application criteria. The classification—and its effects—is the central element of the process. Usually this classification falls within a category, a more general principle of classification. This interaction modifies not only persons as individual but the kinds themselves.
2. People and the classified behaviors.
3. The institutions that surround the topic in question. They comprise established organizations with their own practices or behaviors that affect the habits of the people with whom they interact. It is the institutions that concretize classifications.
4. Knowledge, both specialized and popular.
5. The experts or professionals that generate this knowledge, judge its validity and use it in their practices, who work in the aforementioned institutions, which in their turn guarantee the legitimacy, authenticity and status of the experts.

The interaction between these elements is variable. This outline is but a starting point to carry out singular and specific analyses. The looping effects are complex and depend on very diverse mechanisms. In order to show a general outline in which to inscribe these descriptions, a provisional typology is presented of the processes that produce a looping effect on the kinds of people they are applied to and which, in consequence, change their limits, their characteristics, and are constructed.[19]

[19] In other works, Hacking speaks about the seven "engines of discovery" which emerged and evolved since the beginning of the eighteen century: (1) count, (2) quantify, (3) create norms, (4) correlate, (5) medicalize, (6) biologize, (7) geneticize. These are "searchers" of facts, but also engines to "make up people". There are also engines of practice, such as (8) normalize; of administration, such as (9) bureaucratize; and of resistance to the experts, such as (10) vindication of identity (Hacking 2007a: 305–306).

1. Quantification. The employment of qualities and quantities to classify people and their behaviors. Gradually, qualities tend to become quantities (example: corpulence becomes obesity measured in terms of corporal mass).
2. Biologization. The search for the biological origin of human characteristics, problems, and behaviors (example: trisomy). This biologization of the human kinds does not make them immune to the looping effect.
3. Inaccessible classifications. Some classifications seem inaccessible to the classified persons (example: autism or, as it is now called, autism spectrum disorder)
4. Administrative classifications. Many terms of classifications are used for administrative ends (example: poverty, autism, child abuse). The idea of an administrative kind fits many human kinds, insofar as the sciences that deal with them emerged together with the bureaucratic imperative of distinguishing, enumerating, controlling and improving those that deviate from the norm. Administrative kinds, besides, remind us that there can be a rivalry between administrators, struggles for the same domain of knowledge.
5. Legal classifications: legally refined versions of what Hacking calls common kinds – i.e., classifications of persons that do not have to do with science, such as violent. Criminals are condemned on the basis of legal definitions.
6. Self-appropriation. Of a kind by its members (e.g. gays, prostitutes).
7. Normalization. It has two aspects: the first is the normalization of the kind, which consists in establishing norms for a particular kind, and the second, the normalization of the people classified according to these norms. The idea of the normal is the product of various factors. There are institutions whose aim is to regularize, supervise, enumerate, reform, control, verify, confine, admit in an institution; in a word: normalize.

According to Hacking (1996a) normality is a metaconcept or second-order concept that structures a great amount of ways of thinking. It is of second-order in the sense that it does not apply directly to individual things, but it is accompanied by a substantive sentence, generally a kind of person or behavior. Nothing is normal, *tout court*. The adjective normal has a clear meaning only in conjunction with a nominal term. Moreover, it can be used descriptively to say how things are, but also to say how things should be. What is magical about expression, says Hacking, is that one can frequently do two things with a single expression. The norm might be the usual, but the most powerful ethical constraints are also called norms. One of the reasons for the evaluative load of kinds in the human sciences is that they are generally found in relation to the concept of normal. In fact, Hacking claims that normalization is one of the imperatives of the human sciences, understanding it as an established way of acting that exerts pressure on the research that deals with kinds of people. Hacking's analysis of normality owes a deep debt to Canguilhem's work.

4.6 Looping Effect and Memory

Up to this point I have referred to a public dynamic, but in the looping effect there is also a more private dynamic. The theory and practice of multiple personality are linked to memories and childhood memory, a memory which will not only be recovered but also redescribed. New meanings change the past, which is reinterpreted or even reorganized, repopulated. It is filled with new actions, new intentions, new events that are the cause of the present being. One speaks not only about "*making up people but making up ourselves by reworking our memories*" (Hacking 1995a: 6). It is as if the retroactive redescription changed the past. The action took place, but not under the new description. The past is rewritten because the actions are presented under new descriptions. It might be better to think about past human actions, to a certain extent, as indeterminate. Old actions under new descriptions can be re-experienced in memory. When one remembers what one did, what other people did, it is also possible to rethink, redescribe, and re-feel the past. These redescriptions of the past can be perfectly true. This is why Hacking says that the past is retrospectively revised. This means not only that there is a change of opinion about what was done, but that, as the understanding and the sensibility change, the past becomes full of intentional actions that were not there when it took place. This has happened with the concept of child abuse, which has expanded in such a way that more and more situations fall under its description and there are more and more cases to report. Child abuse leads to arguments following the line that people end up seeing themselves as abusers and/or abused because they are classified in that way. Abuse and the repressed memory of that abuse can have powerful effects on the development of an adult. It is not so much about the truth or falsehood of this proposition, but the sense in which assuming it leads people to redescribe their own past. Individuals explain their experience in a different way and they feel differently about themselves. Many times they can understand their own past in a different way and see themselves—only now—as abused. The events that took place during their life can be seen later as events of a new kind, which was not conceptualized when the experience took place. This is an effect inherent in the concept of child abuse. The concept has its own internal dynamics. In general, knowledge about individuals changes the way in which they think, the possibilities open to them, the kinds of persons and of subjects that "come to be". Knowledge interacts with them and with an extensive body of practices. This generates socially permissible combinations of certain symptoms and illnesses.

4.7 Metaphor of the Ecological Niche

The appearance, mutation and disappearance of transient mental illnesses depends on a space of possibilities –characterized by Hacking as 'ecological niche'[20]— which makes them possible.

Hacking analyzes this topic mainly in his book *Mad Travelers*, which has already pointed out deals with the emergence of transient mental illnesses and a new kind of people, the *fugueur*, through the case of Jean-Albert Dadas (1860–1907, the first *fugueur*). Albert traveled obsessively, frequently without identification papers and sometimes without identity, not knowing who he was and why he traveled. When he 'came to' he had little idea of where he had been, but under hypnosis he was able to recover lost time. Albert's medical reports underscore a small epidemic of compulsive travelers whose epicenter was Bordeaux, in the nineteenth century, but which soon extended to Paris, the whole of France, Italy, and finally to Germany and Russia.

Hacking's interest in this work is not to discuss if this illness is real or constructed. His aim is to provide a framework within which to understand the possibility itself of the transient mental illness (Hacking 1998:1).

What interests him in the case of Albert is not what really happened in his life or what was the cause of his strange behavior. Hacking's concern is about how Albert and his physician, doctor Tissié, established the possibility of the fugue as a proper diagnostic. How they could inaugurate a cataract of *fugueurs* or of diagnostics of fugue. This is not only a question about the fugue, but it is rather the framework within which to think about a whole group of mental illnesses, past and recent, such as anorexia, attention deficit disorder, hyperactivity, autism, among others.

In this search to account for the emergence of the illness, the most important contribution of the text is its metaphor of an ecological niche that allows mental illnesses to prosper. This idea is, according to Hacking, the result of the influence and inspiration of Foucault's (1998: 85–86) linguistic metaphor of discourse or discursive formation. According to the latter, discursive relations explain why in a certain epoch a certain object and/or a certain kind of behavior begins to be spoken about. But why does Hacking not copy Foucault's vocabulary but rather coins another notion? Mainly, because he considers that discourse is not everything. Language is related to the formation of an ecological niche, but also to what people do, how they live and the wider world of the material existence in which they inhabit. This world must be described in all its peculiar and idiosyncratic details, which does not imply that the illness is not real, but rather shows the complex variety of elements that make a new kind of diagnostic possible. The fact that a certain kind of mental illness should appear only in specific historical and geographical contexts does not imply that it is manufactured, artificial, or in some sense not real. What Hacking (1998:

[20] "*I argue that one fruitful idea for understanding transient mental illness is the ecological niche, not just medical, not just coming from the patient, not just from the doctors, but from the concatenation of an extraordinarily large number of diverse types of elements which for a moment provide a stable home for certain types of manifestation of illness*" (Hacking 1998: 13).

16), for his part, is interested in is not discussing whether multiple personality, for instance, is a real disorder, but rather how this configuration of ideas emerges and how they make up and model the lives of certain persons. Rather than understanding reality he aims to provide a framework within which to understand the conditions of possibility of the emergence of these illnesses.

The ecological niche is, according to Hacking, a beneficial idea to understand transient mental illnesses. The ecological niche is not only social and not only medical, it does not provide only from patients or only from doctors; it is the concatenation of an important number of diverse elements that in a certain moment and place provide the space of possibility for certain kinds of manifestations of illnesses.

Through the history of psychiatry two ways of classifying mental illnesses have competed. One model organizes the field according to sets of symptoms, classifying disorders according to how they present themselves. Another organizes it according to their underlying causes. But it is necessary to go beyond symptoms. According to Hacking, in all the natural sciences there is more confidence that something is real when their causes are thought to be understood. Likewise, one feels more confidence when one is skillful in intervening and changing. It is in this sense, as I have remarked, that the questions about multiple personality seem linked to two familiar results in all the sciences: intervention and causation. Hacking remarks that his interest in this topic is obsessively philosophical, because it is about self-reflection. About how a causal understanding, when it becomes known by those who are so understood, can change their character, can change the kind of persons that they are. This can lead to a change in the causal understanding itself. Kinds in the human sciences are formulated with the hope of intervening in the lives of individuals, of changing the basic conditions that can improve people's lives. Causal comprehension is practical (Hacking 1995b: 351).

To analyze the particular and idiosyncratic details mentioned above, Hacking speaks of vectors of different kinds that aim in different directions and suggest the importance of not focusing on a single aspect: discourse, power, suffering, biology.[21] When these vectors are challenged or diverted, the niches are destroyed. Then transient mental illnesses disappear, because they only exist in niches that in some times and places provide a stable site for certain kinds of illnesses. The metaphor of the ecological niche serves as a framework to think about transient mental illnesses, even though this is not the whole story.

Hacking emphasizes four vectors. One is inevitably medical. The illness fits a diagnostic framework, a taxonomy of illnesses. Another is cultural polarity, the good/bad polarity: the illness is situated between two extremes of contemporary culture, one virtuous and the other vicious. What counts as virtue or crime is also a characteristic of society, and virtues are not fixed in time. A vector of observability

[21] Hacking does not understand by vector something technical, but rather uses it as a metaphor. In mechanics, a vectoral force is a force acting in a direction. When there are many vectoral forces acting in different directions, the result is the product of these forces. The metaphor has the virtue of suggesting different kinds of phenomena acting in different senses, whose result can be a possible niche in which a mental illness can prosper.

is also necessary; this is to say that the disorder should be visible as such, as suffering, as something from which to escape. Finally, the illness, in spite of the pain it produces, also provides some liberation that is not available in another part of the culture where it prospers.

Let us analyze now the four vectors, alluding to an example that Hacking develops in *Mad Travelers*: the hysterical fugue as illness, as disorder.

(a) Medical vector

Illnesses are always taxonomized, situated in some nosology, even though the classification can change, as Foucault illustrates in *History of Madness* when he speaks about the spaces where the illness is inserted in the different periods:

> [...] is the manner in which madness was integrated without any apparent difficulty into these new norms of medical theory. The space of classification opens unproblematically to the analysis of madness, and madness immediately finds its place there. (Foucault 2006: 190; 1972a: 208).

When a new illness emerges, the easiest way to give it sense is to host it in some already existing classification. This means, in general, subordinating it to some other superior disorder. In the case of hysterical fugue, the disorders that could act as superior were hysteria and epilepsy. In his works Thomas Kuhn argued that a scientific revolution occurs when a taxonomy of natural kinds must be broken to accommodate a new kind of thing. But a revolution is not needed if the new kind fits within the pre-established order. Hysterical fugue, for instance, could be accepted in the established classification of mental illnesses without any need for a revolution.[22] It did not dislodge existent systems of classification, but it invited a controversy: in what part of the established taxonomy to fit it? This made the illness theoretically interesting for the physicians of the time.

(b) Cultural polarity

Hacking suggests that one of the features of a new mental illness is that it encroaches itself in a bipolar sense in a culture. There are two versions of the same

[22] Hacking claims that when Joseph Babinski, a student of Charcot's demolished the idea of hysteria in force until that moment, insisting that the symptoms were not primarily neurological but a consequence of the suggestion of ideas, and to this was added the influence of new kinds of diagnostics, such as precocious dementia, the miscellany of symptoms of hysteria was redistributed as a group of illnesses. The exit of hysteria from the stage—continues Hacking—dragged with it all its subordinates, among them hysteric fugue. As of that moment, fugue declined because it lacked a vital ingredient in the ecological niche, the medical vector (Hacking 1998: 71–72).

Hacking's claim about the disappearance of hysteria is, to say the least, questionable. Whereas it is certain that the special influence of sociocultural phenomena on the exterior manifestations of hysteria can be pointed out, since its symptoms have varied greatly since Charcot, the same has not happened with the hysteric structure included in the character which, in variable forms, constitutes to this day the permanent and invariable foundation of neurosis. If, as I think, hysteria has not disappeared, it is worth asking in what sense is it a transient mental illness? As well as: is it possible to continue to explain the declination of the hysterical fugue as a consequence of the disappearance of the medical vector as Hacking pretends?

thing, among which the illness insinuates itself: one perceived as virtuous and the other as vicious. The fugue epidemic is produced at a time in which tourism—the virtuous pole—has a formidable development, and vagrancy – the vicious pole—awakens a very strong fear. Both were part of the ecological niche within which this new kind of mental disorder and of experience managed to localize itself. Both make up a pair of elements in the environment that were hosts for the fugue.

Tourism was judged as a beneficial activity. But there was a darker side of travelling, a particularly French obsession with vagrancy. This dark side was also part of the niche. Both were important for the middle classes, because one presupposed leisure, pleasure, whereas the other presupposed a fear of hell. Thus the fugue as a phenomenon was not interesting to common people, who did not make compulsive and senseless trips, for people who could control their fantasies.

(c) Observability

A substantial system of sanitary and police vigilance and detection reported a strange and distorting behavior and contributed to judging it as a mental disorder. The French *fugueurs* were systematically subjected to scrutiny as deserters or draft dodgers. They could not wander across the European continent without being listed by the authorities. For an experience to be judged as a mental disorder, it must be strange, distorting and reported. Hacking refers in this point to Michel Foucault and establishes a difference. The health and police system that determined whether a deserter was precisely that or a sick person was not a strategy conceived in the French philosopher's terms of power/knowledge but it was rather a question of the power of physicians, who considered themselves experts in these cases. The fugue, claims Hacking, is directly related to the systems of social control, without referring to abstractions on power and knowledge, but rather to the police and the military (Hacking 1998: 13).

The hysterical fugue illustrates many social facets of mental illness. It is difficult to say who played a more important role in the definition of a correct diagnostic of the disorder, the patients or the doctors.

(d) Liberation

The fugue was an invitation to escape for a particular kind, for men that had a stable job and certain independence. It was a space for the dysfunctional man, who in that way was able to escape.

If one wishes to grab the nature of a niche, one needs, besides the analysis of vectors, examples of individuals who inhabit it, on the one hand, and on the other different habitats from where the individuals in question are absent. In other words, it is necessary to examine not only positive cases where the fugue is an acceptable medical diagnosis, but also negative cases. For a short time, the hysterical fugue found a niche in France and later in other parts of the European continent. Not so in England and in the United States. Why does the hysterical fugue appear in one place and not in another? What circumstances of the environment made its emergence possible? According to Hacking, the fugue was never taken seriously as a medical entity in the United States. Between 1890 and 1905, problematic persons described with symptoms and experiences that in France and at that time characterized the hysterical fugue, were not presented as *fugueurs* in the United States, and other symptoms were emphasized. The point here is not whether the diagnosis is correct, but to observe that in that country, even though some cases fit the diagnosis of fugue, they were not treated in that way. Only in 1906–7 did French literature cross the Atlantic, but despite the care with which it was discussed, it was singularly irrelevant.

There are reasons why neither in the United States nor in the United Kingdom there is a concern for the fugue. People could disappear there – going to other realms or to distant colonies—much more easily than the French. Migration was not a life choice in France, but it was so in these other two countries. There was an additional reason, and it is that in none of the former were men recruited. Military service had two consequences. First, young men were examined and controlled more rigorously in the European continent that in English-speaking countries. The Napoleonic law of passport was a way of controlling individual movements and in particular to check deserters. Conscription had a second consequence. There was a body of expert forensic physicians that needed to distinguishing deserters by their own will from those who could be excused for medical reasons. In spite of all this, it must be borne in mind that the *fugueurs* of whom we are speaking here were not war deserters; on the contrary, the fugue was a medical entity of times of peace, of boredom and weariness.

In short, hysterical fugue emerges in France but not in England or the United States because in the latter pair the vectors of cultural polarity and observability were not present, and therefore the niche necessary for its emergence was not conformed. Regarding polarity: vagrancy was not a central social problem. Regarding observability: travelers were not systematically inspected through their papers as they were in France due to the search for deserters.

This notion of ecological niche can represent a great contribution to explain the flourishing and disappearance of certain illnesses, in the sense as why the potential dispositions of genetics can become actualized, or why certain illnesses persist when the disorder does not depend on organic abnormalities. In any of these senses, the idea of ecological niche could be a useful and interesting tool, although it is not clear in Hacking's argument whether the notion can be extended – and how—to other disorders or social phenomena that do not have the characteristics of the illnesses he calls transient.

4.8 Different Types of Kinds

Let us now go back to Hacking's consideration that although a notion of kind is not necessary, a discussion on the notion of natural kind could shed light on the relation between human and natural kinds. Why could this be so? Because of his conviction that human kinds are social rather than natural, and that by focusing the attention on the former, new problems are visualized, scarcely treated by philosophers, led him to leaving aside the –in his opinion—idle ontological disputes about social construction and to take seriously the discussion about how different types of kinds are made. All this in order to answer questions such as: how are kinds constructed? Under which restrictions? With what effects? These questions cover all types of kinds: relevant kinds, artefactual kinds, human kinds, and fundamentally the differences between making up kinds of people and making up kinds of things.

In order to answer these questions, and mainly trying to answer the question of whether kinds of people are natural kinds, Hacking undertakes the study of kinds in general. In fact, his articles on natural kinds, as well as elucidating this notion, frequently aim to discuss their relation to the notion of kinds of people.

In his article "Five Parables", when he reflects about the uses of history in philosophy of the natural sciences and the human sciences, Hacking proposes a particular distinction between both kinds of sciences, and mainly a difference between the kinds with which each one of them works, and he shows that history plays a much more important role than merely removing linguistic confusions. This other role is the key to the differentiation between natural and human sciences and the contrast between the creation of phenomena and the making up people.

> Were there perverts before the end of the 19th century?
>
> The answer is NO [...] Perversion was not a disease that lurked about in nature, waiting for a psychiatrist with especially acute powers of observation to discover it hiding everywhere. It was a disease created by a new (functional) understanding of illness (Hacking 1986: 99).

Thus, concludes Hacking, "*[...] we 'make up people' in a stronger sense than we 'make up' the world*" (Hacking 1984: 40). The difference relates to history, since while the objects of the human sciences are constituted by a historical process, those of the natural sciences do not. Even though the latter are created in microsocial conditions, in a determined historical context, once created they become independent from history, they are indifferent to what happens, and in this sense, they are not historically constituted. For this reason, it is claimed that the objects of the human sciences are dynamic, while those of the natural sciences are stable.

As Hacking could ascertain by means of his statistical studies, new ways of counting people are constantly invented, providing new spaces which they can fit and be enumerated. Thus kinds of people and of human actions emerge simultaneously with the invention of the categories that label them, they adapt and interact. Naming practices interact with the things they name. They create new kinds that lead, in part, to new kinds of people. "*We remake the world, but make people*" (Hacking 1984: 49–50). It is this role of history in the practices rather than the words or language what Hacking wishes to underscore. A role that, he believes, is highlighted in the works of Foucault beyond any fascination that words might hold. This role is analyzed in *History of Madness*, by showing how the illness only exists as an object in and by a practice. It has become objectivized in different ways in 25 centuries of western history. This does not mean that social practices have generated something that did not exist in cells or behaviors, or in both, even though behaviors also vary. There is a distinctive substratum that, since the nineteenth century, is considered as a mental illness. At different times it was considered so. But during neoclassicism—even though there were areas in which madness was considered an illness—law practice and general sensibility included it in nonsense. What today is called madness emerged – and disappeared—in different times and sectors, such as divine inspiration, illness, nonsense, celestial punishment or other forms. It never ceased to exist as something that each historical epoch objectivized in a certain way. *History of Madness* shows the impossibility for the unity of a set of statements different by their form and temporally dispersed to be founded on the fact that all of them refer to a single object and that, consequently, what allows us to individualize them should be the referent. A statement is not a signifier, things are not referents. There is no degree zero in the history of madness, a degree zero of the object, a proper essence, a dumb nature that awaits to be discovered and formulated (Castro 1995: 98).

> Fine sport to be sure, but this is not history. Perhaps from one century to another the same name does not refer to the same sicknesses– but this is because fundamentally it is not the same illness that is in question. To speak of madness in the seventeenth and eighteenth century is not, in the strict sense, to speak of 'a sickness of the mind', but of something where both the body and the mind together are in question (Foucault 2006: 214; 1972a: 231–232).

For the formation of archaeological objects it is necessary that a set of determined relations should operate. As Foucault claims in *The Archaeology of Knowledge and the Discourse of Language* (1972b: 45; 1969: 61–62), practices subsist on objects and they, in their turn, are determined by statements, albeit not reduced to them. Things are only drawn in discourse. Objects or things only exist under positive conditions of a complete set of relations. These relations are scattered among institutions, economic and social processes, forms of behavior, systems of norms, administrative measures, techniques, types of classification, modes of characterization, literary expressions, philosophical formulations. The history of an object cannot be but the history of what has been said about it.

This is why Hacking considers that the works of the French philosopher are useful to grasp the interrelations of power and knowledge that constitute human beings as such. This role played by history in the constitution of persons through practices is the way in which it can exert the strongest impact on philosophy.

This distinctive role played by history in the human sciences has a consequence, for instance, that the main difference between sciences such as physics and psychiatry, says Hacking, is about their effects; to the feedback effect of new classifications, peculiar of the human sciences. The introduction of the kind 'pulsar' allows astrophysicists to see a great deal of pulsars. The introduction of new kinds creates a new world in which the scientist works. However, the number of pulsars continues to be the same, beyond the fact that the world with which the astrophysicist works has changed. For its part, the introduction of kinds such as multiple personality, according to Hacking, not only changes the world in which psychiatrists work but it also changes the world, insofar as from that moment on many more people will exhibit the disorder. The creation of the new kind has come to allow for and reinforce a new way of behaving. This is the difference that Hacking points out between classes with which natural sciences work and those which human sciences work. In this regard, Hacking has remarked that the difference he proposes is not related to the proposal that the natural sciences look to explicate whereas the social sciences look to understand, nor with that according to which the latter are constructed whereas the former are not. He does not deny that these differences exist, but they are not the ones he proposes.

The effect of classifications and how new kinds are created in the human sciences is related to a whole range of disciplines and subdisciplines that include from labelling to the cognitive sciences. However, there is not an inherited conception of the philosophical study of the classification of people as there is a philosophical tradition related to the classifications of natural objects: the doctrine of natural kinds, in Bertrand Russell's terms. Few in this tradition pay attention to how the very word 'nature', apparently so transparent, is polemical, ideological, and framed in a theory of places, roles and duties of human beings and their situation in the world.

4.9 Kinds of People

Due to his analytical background, Hacking started thinking about kinds of people in terms of the natural kinds. In 1991 he claims that a historical perspective allows him to recognize the existence of more than one kind of natural kind, each one of which has a distinctive history, in a clear allusion to the different history of kinds of people (Hacking 1991a: 110). But the presence of the particular phenomena within the latter soon led Hacking to start referring to them as human kinds, to distinguish them from natural kinds.

In 1995, Hacking (1995b: 352–354) characterizes human kinds as those which are relevant to some of us; that classify people, actions and behaviors; they are studied in the human and social sciences and they include behaviors, actions or tendencies only when they are projected to form the idea of a kind of person. They are kinds about which we would like to have a systematic, general and precise knowledge; classifications that could be used to formulate general truths about people; generalizations strong enough to look like laws about them, their actions or their feelings. Laws strong enough to predict what individuals will do or how they will respond to the attempts to help them or modify their behavior. The difference Hacking will propose between human and natural kinds is not related to a hermeneutic or constructivist stance about human kinds versus a causalist stance about natural kinds. On the contrary, in both types of sciences only one kind of causality is considered relevant: efficient causality.

After a long time, Hacking understood that his notion of human kinds was confused, and he abandoned it. Even then, he continued to accept some idea of natural kind, which led him to conceive a defined kind of kinds which he called interactive, in opposition to indifferent kinds (Hacking 2007a: 293, footnote 21). These new concepts: indifferent and interactive, mark the fundamental difference between the kinds of the natural and the human sciences, respectively. In indifferent/natural kinds there is no looping effect; in interactive/human kinds there is. In the former, their members are indifferent to the classification. Indifferent, because even though our knowledge of them affects them and interact with the way in which we intervene, they have no knowledge of how they are classified. In the latter, according to Hacking, the classified individuals interact with the classification and its criteria of application, with the institutions that surround the topic in question, with specialized and popular knowledge about it, etc., and they know how they are being classified.

The concept of interactive has, for Hacking, the merit of focusing on the actor, agency and action. The prefix inter- suggests the way in which the classification and the classified individual can interact, the way in which actors are aware that they are classified in a certain way, if only by being treated or institutionalized in a certain way, and so they experience themselves in this way. Hacking's classifications – like Foucault's words—do not exist in the vacuum of language. They exist in the midst of a social and material matrix. Names do not work alone. They work in the context of practices in general, of institutions, of the material practices of things and of the other people. An analysis of the classifications of human beings is an analysis of classificatory words in the sites and situations in which they are used. This is not so for indifferent classes. Phosphorus does not change because it is labelled in this way. It is true that insofar as we know it we will be able to use it, but that we should describe it as phosphorus will not make any difference to it. Phosphorous does not know itself as such, and in this sense, it does not react to the classification consciously. Very different is, according to Hacking, what happens when we classify someone as a homosexual. Being called a homosexual or thinking about oneself as such changes the way of being, the behaviors, the choices of that individual. Therefore, to admit a new interactive kind[23] produces a particular feedback effect in the world, as it results in new ways of being and behaving for its members.

But, as I have already pointed out, in 2007 Hacking claims that there are no natural kinds. The collapse of this idea led him to reject definitively the idea of human kinds as opposed to natural kinds. If there is no such thing as a defined type of kinds that could be called 'natural', there cannot be any other kinds, such as human kinds, that are defined by opposition to them. There are no two types of kinds that could be defined in a sufficiently clear way so as to be distinguished one from the other.

There would be, however, in my view, an interesting proposal to visualize Hacking's stance, which, even though it does not rule out natural kinds it does not understand them in an essentialist way either. I refer to the promiscuous realism proposed by John Dupré (1993), who also claims that there is no single classification, complete and exhaustive. The vast and complex structure of the world can be categorized in different forms that crisscross each other and that respond to the different specifications of the objectives that underlie these attempts of classification. There are many possible and defensible ways of classifying, and which should be

[23] In general, Hacking speaks of new kinds, but it is necessary to point out that they might be kinds that have old names but have acquired new meanings in the light of a new knowledge.

the best one will depend on the purposes of the classification and the peculiarities of the individual in question. In this sense, I understand that there would be no reason to abandon the distinction between natural and human kinds, that it could coexist with that between interactive and indifferent kinds, since Hacking, besides, does not present them as exclusive. A single entity, moreover, can be a member of different classes. This could be the case with the question Hacking poses in *The Social Construction of What?* about there being a deep-rooted conviction that retarded children, the schizophrenic, and autistic people suffer neurological or biochemical problems that will be identified in the future, and that in that case they could belong at the same time to a natural and an interactive kind. It could be the case of Down's syndrome, caused by a trisomy on chromosome 21, and that at the same time it can be considered an interactive kind. The way in which we choose to classify an individual depends on whether we focus on the pathology or its interaction with the classification. In any case, neither classification is privileged in relation to the other. This classificatory pluralism does not mean abandoning realism, and it agrees with Hacking's idea that the world is so complex that there is no humanly accessible theory about the whole. No theory can offer more than a perspective of the world.

In abandoning not only the use of the notion of human kinds but also the idea that there exists such a type of kinds, Hacking also abandons the notion of interactive kind, and therefore, also indifferent kinds, which was its opposite. He continues to vindicate the interaction between classifications, people, institutions, knowledge and experts, as essential elements for the explanation of the looping effect and making up people, but he does not accept that there is a well-defined type of classification of persons equivalent to what he called interactive or human kinds. "*Interaction, yes, but interactive kinds as a distinct class, no*" (Hacking 2007a: 293, footnote 21).

This statement concludes, in a way, a process Hacking started in "Five Parables" when he pretended to present an old distinction between natural and human sciences from a novel point of view, and he centered that difference in the existence or not of the looping effect between the classifications and the classified individuals. There emerged, immediately, the question about what was the difference Hacking pointed out between the looping effect in the human kinds and that which can take place in at least some of the classes with which the natural sciences work, for instance in biology, where there are microorganisms, morphological characteristics of certain plants, etc., which change as a reaction to how we classify them. As the experts know them, classify them, submit them to certain practices and expect them to behave in a certain way, some of them modify themselves to resist, causing a change in the kind and what is known about it. Hacking emphasized then that the difference should not be grounded only on the fact that the looping effect takes place, but rather on the fact that in the human sciences this occurs in a particular sense, because the subjects become aware of the way in which they are classified. When human beings know the classifications that affect them, they can change the experience they have of themselves, making what was known until them incorrect.

In this respect, in *The Social Construction of What?* Hacking chose the kind of hyperactive children to illustrate this phenomenon, insisting that "*children are conscious, self-conscious, very aware of their social environment, less articulate than many adults, perhaps, but, in a word, aware*" (1999a: 103), unlike what happens in the kinds of natural sciences, where "*calling a quark a quark makes no difference to the quark*" (Ibid: 105). It should be clear, then, that when Hacking spoke of the interactive kinds of the human sciences, he was not referring to the mere interaction between classification and the members of the kind—as seemed to be the case in the original formulation—but rather that the interaction that could establish a difference between the natural and human sciences was the interaction with individuals that know how they are classified, since an interaction – even without awareness on the part of the classified being—can also take place with the objects of the natural sciences. However, whereas in different passages of the aforementioned work Hacking alludes to this, he does not always achieve that clarity due to the fact that what he generally emphasizes in the distinction is the interaction of classification, leaving in a second plane the idea that such interaction is distinguished by being conscious.

This happens in important passages such as the following:

> [...] a cardinal difference between the traditional natural and social sciences is that the classifications employed in the natural sciences are indifferent kinds, while those employed in the social sciences are mostly interactive kinds. The targets of the natural sciences are stationary. Because of looping effects, the targets of the social sciences are on the move (Hacking 1999a: 108).

In the first place, I reiterate that what is substantial is not the classification, as this passage seems to suggest, but rather the different kinds of objects that react in different ways to the classification. Now, if this is so, the pretended novelty of Hacking's position is damaged, since it is permissible to ask about the familiarity of this distinction with other proposals defended from the hermeneutic conception.[24] In the second place, being in constant movement is not what makes the difference, but, in any case, the kind of movement produced due to the fact that human beings know the classification. If the ontological difference of the objects is not mentioned, not only is the distinction between the sciences not stressed, but it becomes impossible even to sustain the difference between classifications, because the idea of interaction in itself does not illustrate the distinction that it is supposed to ground. On the contrary, what difference would be between what Hacking remarks about autistic people:

[24] Let us recall, for instance, what A. Giddens said a few years earlier: "*The concepts and theories produced in the natural sciences quite regularly filter into lay discourse and become appropriated as elements of everyday frames of reference. But this is of no relevance, of course, to the world of nature itself; whereas the appropriation of technical concepts and theories invented by the social scientists can turn them into constituting elements of that very 'subject matter' they were coined to characterize, arid by that token alter the context of their application*". A. Giddens cited in Knorr Cetina (1981: 146). For his part, John Searle, in *Minds, brains and science* (1984: 74) –the same year Hacking published "Five Parables"—posed the distinction between both kinds of sciences as follows: "*What are the fundamental principles on which we categorize psychological and social phenomena? One crucial feature is this: for a large number of social and psychological phenomena the concept that names the phenomenon is itself a constituent of the phenomenon. [...] many of the terms that describe social phenomena have to enter into their constitution. And this has the further result that such terms have a peculiar kind of self-referentiality*".

> [...] by interaction I do not mean only the self-conscious reaction of a single individual to how she is classified. I mean the consequences of being so classified for the whole class of individuals and other people with whom they are intimately connected (Hacking 1999a: 115).

> And about microbes?
> Microbes, not individually but as a class, may well interact with the way in which we intervene in the life of microbes. We try to kill bad microbes with penicillin derivatives. We cultivate good ones such as the acidophilus and bifidus we grow to make yogurt. In evolutionary terms, it is very good for these benevolent organisms that we like yogurt, and cultivate them (Hacking 1999a: 105).

Once again, the fundamental difference is the awareness that autistic people and those around them supposedly have about classification. I claim that it is the awareness they supposedly have, because the reproduced fragment introduces us to another interesting question, what does Hacking refer to when he speaks of knowledge or awareness that people have about how they are classified? I am always surprised by statements such as the following: "[child viewers] *are well aware of theories about the child viewer and adapt to, react against, or reject them*" (Hacking 1999a: 27).

Regarding autistic people, even though he recognizes how problematic it is to give this example for his idea of interactive kind, he also affirms that they in their way know, are aware, etc. It is hard to understand this choice of Hacking's, since the experts point out in general the incapacity of at least some of these children to establish adequate systems of communication with their environment, even on occasions when there is no communication and the relationship with the rest of the people is null. The experts also point out the rupture with reality as a consequence of the non-precise delimitation of the environment in itself, as well as the characteristic of implementing types of construction of representations of the world different from the usual. Taking these features of the autistic child as a starting point, it is hard to see how the idea that they are aware and have knowledge of how they are classified should be interpreted.

Hacking has pointed out that there is a type of kinds – which include autism—he calls inaccessible, characterized precisely because its members cannot assimilate how they are being classified. According to Hacking (1995b: 374) in them there cannot be self-conscious feedback. However, there can be a loop involving a larger human unit, for example, the family. A bureaucratic world constituted by pedagogues, psychologists, educators, to which autistic children are integrated in institutional practices. Interactions and looping effects would take place, in this way, at the institutional level of pedagogical, psychological, educational bureaucracy.

What has been pointed out until now reinforces the conclusion that the idea of indifferent kinds versus interactive kinds cannot make the distinction between natural and human sciences in the manner Hacking pretended at the beginning. However, and in spite of this, I believe that a good part of the theory generated on this subject continues to be of great value. In the first place, the fundamentally illustrative aspect of his notion of looping effect to express the impact of classifications on human beings. It is known that the sciences have cultural effects, but Hacking shows these effects not in a general sense, as it is usually done, acting on the lives of individuals qua society, but on a particular level: he shows how the individual human being is affected by what science says about them and how it says these things. Secondly, it allows us to think that there are different types of kinds according to the relation constituted between the classification and the classified members, so that we can speak of kinds in which there is no interaction at all, kinds in which there is some kind of interaction even though not as a consequence of the knowledge the members of the class have of being classified in a certain way, and finally, others in which the interaction is the result of a self-aware feedback. To think about it in this manner not only goes in Hacking's direction that the world is too complex to make it fit in a single classification, but it also enhances the ways of thinking and doing in the sciences.

References

Bird, A. (2010). *The metaphysics of natural kinds*. http://eis.bris.ac.uk/~plajb/research/inprogress/Metaphysics_Natural_Kinds.pdf

Castro, E. (1995). *Pensar a Foucault. Interrogantes filosóficos de La arqueología del saber*. Buenos Aires: Biblos.

Díaz, E. (2003). *La filosofía de Michel Foucault*. (2ª. ed.) Buenos Aires: Biblos.

Dupré, J. (1993). *The disorder of things. Metaphysical foundations of the disunity of science*. Cambridge: Harvard University.

Eco, U. (1999). *Kant and the platypus: Essays on language and cognition*. New York: HMH.

Foucault, M. (1969). *L'archéologie du savoir*. Paris: Gallimard.

Foucault, M. (1972a). *Histoire de la folie à l'âge classique*. Paris: Gallimard.

Foucault, M. (1972b). *The archaeology of knowledge and the discourse of language*. New York: Pantheon Books.

Foucault, M. (1980). *Power/knowledge: Selected interviews & other writings 1972–1977*. New York: Pantheon Books.

Foucault, M. (1984a). ¿What Is Enlightenment?. In P. Rabinow, *The Foucault reader* (pp. 32–50). New York: Pantheon Books.

Foucault, M. (1984b). L'éthique du souci de soi comme pratique de la liberté. In M. Foucault (1994), *Dits et écrits* (Vols. 1–4, pp. 1527–1539). Paris: Gallimard, édités par D. Deferí & F. Ewald.

Foucault, M. (1984c). Foucault. En M. Foucault (1994), *Dits et écrits* (Vols. 1–4, pp. 1450–1454). Paris: Gallimard, édités par D. Deferí & F. Ewald.

Foucault, M. (1998). *Aesthetics, method, and epistemology. Essential works of Foucault 1954–1984* (Vol. 2). New York: The New Press.

Foucault, M. (2000). *Ethics, subjectivity and truth. Essential works of Foucault 1954–1984* (Vol. 1). New York: Penguin Books.

Foucault, M. (2006). *History of madness*. London and New York: Routledge.

Hacking, I. (1981). The archaeology of Michel Foucault. In I. Hacking (2002), *Historical ontology* (pp. 73–86). London: Harvard University.

Hacking, I. (1983). *Representing and intervening*. Cambridge: Cambridge University.

Hacking, I. (1984). Five parables. In I. Hacking (2002), *Historical ontology* (pp. 27–50). London: Harvard University.

Hacking, I. (1986). Making up people. In I. Hacking (2002), *Historical ontology* (pp. 99–114). London: Harvard University.

Hacking, I. (1990). Natural kinds. In R. B. Barrett, & R. F. Gibson (1990), *Perspectives on Quine* (pp. 129–141). Oxford: Blackwell.

Hacking, I. (1991a). A tradition of natural kinds. *Philosophical Studies, 61*(1–2), 109–126.

Hacking, I. (1991b). The making and molding of child abuse. *Critical Inquiry, 17*, 253–288.

Hacking, I. (1993). Working in a new world: The taxonomic solution. In P. Horwich (Ed.), *World changes. Thomas Kuhn and the nature of science* (pp. 275–309). Cambridge: MIT.

Hacking, I. (1995a). *Rewriting the soul. Multiple personality and the sciences of memory*. Princeton: Princeton University.

Hacking, I. (1995b). The looping effects of human kinds. In D. Sperber, D. Premack, & A. J. Premack (Eds.), *Causal cognition: A multi-disciplinary debate* (pp. 351–383). New York: Oxford University.

Hacking, I. (1996a). Normal people. In D. R. Olson & N. Torrance (Eds.), *Modes of thought. Explorations in culture and cognition* (pp. 59–71). Cambridge: Cambridge University.

Hacking, I. (1996b). The disunities of the sciences. In P. Galison & D. Stump (Eds.), *The disunity of science. Boundaries, contexts and power* (pp. 37–74). Stanford: Stanford University.

Hacking, I. (1998). *Mad travelers. Reflections on the reality of transient mental illnesses*. Charlottesville: University of Virginia.

Hacking, I. (1999a). *The social construction of what?* Cambridge: Harvard University.

Hacking, I. (1999b). Historical ontology. In I. Hacking (2002a), *Historical ontology* (pp. 1–26). London: Harvard University.

Hacking, I. (2001). *An introduction to probability and inductive logic*. Cambridge: Cambridge University.

Hacking, I. (2002a). *Historical ontology*. London: Harvard University.

Hacking, I. (2002b). How 'natural' are 'kinds' of sexual orientation? *Law and Philosophy, 21*(3), 335–347.

Hacking, I. (2007a). Kinds of people: Moving targets. *Proceedings of the British Academy, 151*(p), 285–318.

Hacking, I. (2007b). Natural kinds: Rosy Dawn, scholastic twilight. In *Royal Institute of Philosophy supplement* (pp. 203–239). Cambridge: Cambridge University.

Knorr Cetina, K. (1981). *The manufacture of knowledge. An essay on constructivist and contextual nature of science*. Oxford: Pergamon Press.

Kuhn, T. (2000). *The road since structure*. Chicago: University of Chicago.

Lewowicz, L. (2005). *Del relativismo lingüístico al relativismo ontológico en el último Kuhn*. Montevideo: Facultad de Humanidades y Ciencias de la Educación.

Nietzsche, F. (2001). *The gay science*. Cambridge: Cambridge University.

Quine, W. V. O. (1969). Natural kinds. In W. V. O. Quine, *Ontological relativity and other essays* (pp. 114–138). New York: Columbia University.

Regner, A. C. (2000). Conversando con Ian Hacking. *Episteme, 10*, 9–16.

Searle, J. (1984). *Minds, brains and science*. Cambridge: Harvard University.

Vagelli, M. (2014). Ian Hacking. The philosopher of the present. *Iride, 27*(72), 239 269.

Veyne, P. (2009). *Foucault. Pensamiento y vida*. Barcelona: Paidós.

Chapter 5
Classifications, Looping Effect and Power

> *[…] a new body of knowledge brings into being a new class of people or institutions that can exercise a new kind of power*
> Hacking (1981: 73).

Abstract Even though Ian Hacking vindicates in general the influence of Michel Foucault's the archaeological stage on his own thought, I posit that in many aspects of his proposal for the human sciences there is an imprint of the genealogical stage of the French philosopher—although Hacking does not deal with power, one of the concepts that characterize Foucault's genealogy, at least not in a systematic and explicit way. In this chapter, *Hacking and Foucault. Classifications, looping effect and power*, I discuss the fundamental role of power in relation with the notions of classification, looping effect, and making up people that Hacking proposes for the human sciences. To this end, I start with a brief outline of the idea of power in Foucault, in order to show how the elements he takes as essential in a power relation can be clearly identified in the looping effect. I illustrate how power is imbricated in Hacking's proposal, resorting to several of the examples he himself has used to develop the notions mentioned in the title of this chapter.

Keywords Classifications · Looping effect · Power and resistance in Michel Foucault · Genealogy · Ian Hacking

Before continuing with the next node, I think this is an opportune moment to stop at a point that I have already mentioned. Hacking vindicates in general the influence of the archaeological stage of Foucauldian thought on his own thought. However, as I have already remarked, I believe that in many aspects there is an imprint of the genealogical stage of the French philosopher. In spite of this, Hacking does not work on one of the concepts that characterize Foucault's genealogy: power. At least,

M. L. Martínez Rodríguez, *Texture in the Work of Ian Hacking*, Synthese Library 435, https://doi.org/10.1007/978-3-030-64785-8_5

he does not do so in a systematic and explicit way. However, this does not mean that he denies the power-knowledge relation established by Foucault. On the contrary, not only does he not deny it but he admits that Foucault's works are useful to capture interrelations of power and knowledge that literally constitute us as human beings (Hacking 1984: 50). I believe that, in spite of the fact that Hacking does not deal with the aforementioned notion of power in the specified sense, it is still possible to see power playing a central role in relation with his notions of classification, looping effect and making up people in the human sciences. In this chapter, I will show how I see the Foucauldian notion of power working in Hacking's proposal. To this end, I will start by outlining the idea of power of the French philosopher.

5.1 An Overview of Power in Foucault

Foucault's concern with power is not related to legitimacy but rather to its workings, with the specificity of its mechanisms, the perception of its relations, its extensions. To describe it, Foucault undertakes on the one hand an explanation of which categories must be left aside and why, and on the other the elaboration of appropriate conceptual tools to describe the logic characteristic of power relation and the struggles established around them, a research can only be carried out gradually, departing from specific situations.

Foucault does not like the term "theory" to refer to his work on power. In *The History of Sexuality. Volume I* (1976: 109, 1978: 82), he remarks that the aim of his investigation "*is to move less toward a 'theory' of power than toward an 'analytics' of powe*r". And in *Security, Territory, Population* (2009), he claims that "*this analysis simply involves investigating where and how, between whom, between what points, according to what processes, and with what effects, power is applied*" (2004: 2, 2009: 2).

If the objective were to build a theory of power, it would be necessary to consider it as emerging in a given point and moment; it would be necessary to undertake its genesis and then its deduction. But if power is understood as an open bundle, more or less coordinated, of relations, then the main problem is to provide a grid for its analysis that should allow for an analytics of power relations (Foucault 2008: 186). In other words, it is necessary to work "*toward a definition of the specific domain formed by relations of power, and toward a determination of the instruments that will make possible its analysis*" (Foucault 1976: 109, 1978: 82).

The analytics of power establishes the mutual conditioning between knowledge techniques and power strategies: power relations open fields of objects, they allow for research procedures and they configure subjects of knowledge; the knowledge and effects of truth that, for their part, involve, support and prolong the reach of those strategies. The proliferation of dispositives of power-knowledge has the shape of an ascending spiral; among them there is a play of mutual supposition and feedback that makes one to be the condition for the sustaining and strengthening of the other. There is no exteriority between power and knowledge, neither is there confusion, but an articulation that departs from its difference. They have a correlative relation, determined in each historical specificity.

Foucault does not develop an exposition of the principles of power because somehow his thought as a whole consists in saying that power has no principles. Nevertheless, he questions a certain number of postulates about it. They are not about principles that appear in a specific theory, but rather about implicit postulates that, in his view, run through theories of power as a whole. But Foucault is not content with saying that it is necessary to rethink some notions; he proposes new coordinates for practice that run counter to the questioned postulates.

1. Postulate of property. Power is a property of something or someone, of a class for instance, that would have conquered it.

On the contrary, power, according to Foucault, is not a property but a strategy that "*defines innumerable points of confrontation, focuses on instability, each of which has its own risks of conflict, of struggles, and of an at least temporary inversion of the power relations*" (Foucault 1975: 32, 1995: 27). Its effects are not attributable to an appropriation but to dispositions, maneuvers, tactics, techniques, workings. Power is understood from a functionalist perspective; it is exercised rather than possessed; it is not the privilege acquired or preserved of the dominant class, but the effect of the whole of its strategic positions. It lacks homogeneity, it is defined by singularities, the singular points it passes through. It is exercised across the whole surface of the social field, according to a system of turnovers, connections, leverage points, things as tenuous as the family, sexual relations, housing, etc. (Foucault 2013: 231, 2015: 228).

2. Postulate of localization. Power is localized in the State apparatus.

On the contrary, Foucault claims that power is not localized in the State apparatus, but that the State apparatus is an overall effect or the result of a multiplicity of gears and nuclei situated on a completely different level, and which constitute in themselves a microphysics of power. The State presupposes power relations but does not explain them. It presupposes the relations of strength that come from another part. Whereas power is local, in the sense that it is never global, always consisting of local foci, it is not local in the sense of localizable because it is diffuse, it is dispersed across the whole social field. Power relations work under the grand systems.

In order to understand what power is, one must not take as a starting point the large, molar entities; rather, it is a question of moving from a macro to a micro dimension. Power must be captured on the level of molecules and corpuscles, of the quotidian, from individual, specific relations of strength. What has to be shown is its genesis and behavior. The State, classes, the law, are molar powers that actualize, that integrate the molecular relations of strength.

3. Postulate of subordination. Power embodied in the State apparatus is subordinated to a mode of production as infrastructure.

Far from this, Foucault believes, power is immanent. One should not speak about relations of production without introducing power relations and without intertwining both. Power relations are not in a position of exteriority with respect to other kinds of relations; they are not in the position of superstructure; they are where they play a directly productive role (Foucault 2013: 234, 2015: 231).

4. Postulate of essence or attribute. Power has an essence and is an attribute that qualifies those who possess it, distinguishing them from those over whom it is exercised.

Antagonistically, for Foucault power lacks essence. It is not an attribute that qualifies those who possess it—the dominators—distinguishing them from those over whom it is exercised—the dominated. Power is functional in operative terms. It is not an attribute but a relation: the relation of power is the set of relations of force that circulates both through the dominated forces (the affected point) and through the dominant ones (affecting point): both constitute singularities. This means that power is not a form and the relation of power is not produced between two forms. We must speak about a relation of power in the singular, and a relation of forces in the plural. "Force" does not exist in the singular. Every force presupposes a relation with another force. The power relation is the relation of force with force, an action over another action, real or possible; it is implanted where those forces exist, even if they are minuscule (Foucault 1975: 31–32, 1995: 26–27, 2013: 236, 2015: 233). This means that force is fundamentally an element within a multiplicity. A force relation is a function of the kind of inciting, stirring, combining, dissuading, facilitating, widening, imitating. In *Discipline and Punish*, Foucault speaks about a first rank of function (or categories of power), to organize in space: to distribute in space, to enclose, serialize, align, grid. A second function, to order in time: to subdivide one's own time, to program an event, to decompose a gesture. At work, in the workshop, in the factory. A third function, to compose in space-time: to produce a useful effect superior to the sum total of elemental forces, to constitute a productive force whose effect must be superior to the sum total of the elemental forces that compose it.

Normalization is the force relation par excellence; it consists in distributing in space, ordering in time, composing in space-time. Modern societies do not proceed by means of ideology or repression, but by normalization.

5. Postulate of modality. Power acts through violence or ideology. Violence is the relation of a force on a thing, object or being; it is exercises on the support of an action, on the subject of an action. So understood, power only represses.

According to Foucault, power effects must not always be described in negative terms: it excludes, represses, rejects, censors, abstracts, conceals, hides. In fact, power, rather than repressing, produces reality, and rather than ideologizing, abstracting or concealing, produces truth. *Nietzsche, Genealogy, History* (1980b), is the milestone that marks the passage from Foucault's repressive hypothesis about power to this other, positive one. Foucault considers the notion of repression completely inadequate to account for what there is of productivity in power, of what there is of utility in it. If power was nothing but repressive, if it did nothing else apart from saying *no*, would it really be obeyed? What makes power stick, what makes it be accepted, is simply that it does not only function as a force that says no, but that it produces things, it induces pleasure, it shapes knowledge, it produces discourses; it is necessary to consider it as a productive network that encompasses the whole social body rather than as a negative instance whose function is to repress (Foucault 1980a: 119).

It is in this sense that Foucault denies that modern power is primarily exercised by means of repression and that the opposition to repression is an effective way of resisting it. He believes that modern power, for instance, created new forms of sexuality by inventing discourses about it. Even though sexual relations themselves took place throughout human history, homosexuality as a distinct category, with defined psychological, physiological and perhaps genetic features, was created by the system of power/knowledge of the modern sciences of sexuality. This organizational aspect of power is what makes Foucault say: power produces the Real.

6. Postulate of legality.
   Foucault denies the link that most theories establish between the State and the law.

After this brief overview of Foucault's questionings of the traditional postulates of power, it is time to point out that for him what defines power relations is a form of action that does not act directly and immediately on men, but rather on their actions, existent actions or others that might emerge in the present and the future. They can only be articulated on the basis of two indispensable elements: 1—the "other" must be entirely recognized and sustained until the end as a person that acts, and 2—in the face of a power relation, a whole field of replies, reactions, results and possible interventions opens up. Power is a tensioned arc between a force and a resistance.

With regards to the first element, according to Foucault, when one wishes to define a category of power what matters is not the objects or beings it is applied to, since the category of power in itself is the relation of force with other forces and not with objects or beings. The object or being is a shaped material. The category of power is trans-qualitative. It runs through qualities, keeping only unshaped, non-qualified material. The example par excellence of the category of power is to impose any kind of task whatsoever to any human multiplicity. Of course, in actual fact this is inseparable from the categories of knowledge, that is to say, of shaped material. Knowledge and power, as has been said, form a concrete, mixed group. One cannot speak about power in a pure state, only about power already embodied in a knowledge.

Power is actualized in knowledge; the strategic is actualized in the strata. What does it mean to actualize? To integrate, to differentiate. To fix the matter, shape it and finalize the function. Knowledge concerns historical stratified formations, shaped matters and formalized functions, distributed segment by segment, under the two great formal conditions: the visible and the sayable. Power, on the other hand, is diagrammatic. It mobilizes non-stratified matter and functions, it utilizes a very flexible segmentarity. It does not run through forms but rather through singular points that always indicate the application of a force, the action or reaction of a force in relation to others. Power is informal, knowledge is organization of forms. There are forms of knowing and points of power.

Knowledge and power are irreducible practices, of different nature, but between them there is a reciprocal presupposition and a mutual capture. The institution constitutes the inevitable factor of integration between both, where the relations of force are articulated in forms: forms of visibility, such as institutional apparatuses, and forms of enunciability, such as their rules. The institution is the eminent place where the exercise of power is a condition of possibility of knowledge, and where the exercise of knowledge becomes an instrument of power. The institution is the point of encounter between stratum and strategy, where the archive of knowledge and the diagram of power are blended and interpenetrate each other, without becoming confused. It is not the institution that explains power but the other way around, it is power that explains the institution. The role of the institution is to give power the means to reproduce itself.

Regarding the second element, unlike a relation of violence that can only be actualized by immobilizing and reifying the individual, power presupposes that the subject be at all times recognized as a person who acts and who, therefore, unfolds with its exercise a whole open space of possible answers. Power presupposes freedom. Power is exercised only on free subjects that face a field of possibilities where they can develop several forms of behavior, several reactions and several behaviors. Hence its field of application is possible actions. Power,

> […] is a total structure of actions brought to bear upon possible actions; it incites, it induces, it seduces, it makes easier or more difficult; in the extreme it constrains or forbids absolutely; it is nevertheless always a way of acting upon an acting subject or acting subjects by virtue of their acting or being capable of action (Foucault 1982: 789).

This implies a transformation of the acting subjects; a transformation in the shape of a choice, reaction or behavior. Resistance is, for subjects invested with power, the possibility (but also the need) to transform themselves and to react.

There is no mutually exclusive face-to-face confrontation between power and freedom, but rather a much more complex interrelation. In this game, freedom can appear as the condition for the exercise of power. For this reason, the relations between power and freedom's denial to submit cannot be separated. The crucial problem of power is not that of voluntary servitude. The true center of power relations is the reluctance of will and the intransigence of freedom, a relation that is at the same time incitation and struggle; more a permanent provocation than a face-to-face confrontation that paralyzes both sides.

5.2 Classification, Looping Effect and Power

After this brief introduction to the idea of power in Foucault and in relation to the title of this item, it is necessary to point out that Hacking, in general, does not refer to the power of classifications. However, I consider that it is difficult to deny that the classifications of the human sciences, when they interact—according to Hacking—with that which they classify, exhibit what we have seen is the specific nature of power according to Foucault: the exercise of power is not simply a relation between interlocutors, individual or collective; it is a way in which certain actions modify others.

In my view, power is present in the experts that in *Rewriting the Soul* contribute to the emergence of multiple personality departing from institutionalized forms of knowledge about these kinds of people, and that they map a complex interactive field between systems of knowledge, society, and persons affected. It is present in the institutions and experts that label the *fugueurs* of *Mad Travelers* and in the consequent resistance of the patients so classified. It is present in the experts of *The Taming of Chance*, who every now and then ideate and organize new kinds of people to which a group of individuals adjust in each new census, and to which they adapt or react, taking power, and forcing these experts to modify the classification adjusting it to the new reality.

According to Hacking, the experts tend to dominate what is known, they have the power to affect the members of the kind—Foucault's words—and the latter, in their turn, have the power of being affected and tend to react to the senses in which the experts expect them to act. This is to say, Hacking's idea of looping effect somehow presupposes, as I pointed out in a previous section with regards to the French philosopher, that 1- the person classified is recognized by the expert and sustained until the end as a person that acts, that reacts, and 2- faced with this classification, for the person classified a whole new field of responses, reactions, results and possible interventions opens up, which will finally result in the formation of the looping effect.

In other words, the two elements that Foucault points out as indispensable for a power relation become clearly visible, in my opinion, in the notion of looping effect that Hacking proposes to describe the relation between the classifications in the field of the human sciences, in the interaction between the categorization and the persons categorized.

It is true that in Hacking there does not appear an explicit thematization of the intentionality of the classifications and their consequences, such as it appears in the works of Foucault when he claims, for example in *Discipline and Punish*:

> The first of the great operations of discipline is, therefore, the constitution of *'tableaux vivants'*, which transform the confused, useless or dangerous multitudes into ordered multiplicities. The drawing up of 'tables' was one of the great problems of the scientific, political and economic technology of the eighteenth century: how one was to arrange botanical and zoological gardens, and construct at the same time rational classifications of living beings; how one was to observe, supervise, regularize the circulation of commodities and money and thus build up an economic table that might serve as the principle of the

> increase of wealth; how one was to inspect men, observe their presence and absence and constitute a general and permanent register of the armed forces; how one was to distribute patients, separate them from one another, divide up the hospital space and make a systematic classification of diseases: these were all twin operations in which the two elements—distribution and analysis, supervision and intelligibility—are inextricably bound up (1975: 149–150, 1995: 148).

It is also fair to point out that Hacking, unlike Foucault, does not speak expressly about organizing in space, serializing, alienating, gridding, etc., as a first rank or category of power. But this does not prevent him from dealing with and working on situations of this kind. Not only in *The Taming of Chance*, which would be the most evident text given its subject matter, but also in other texts and articles, and not merely in general terms, but by means of concrete examples.

In *The Taming of Chance*, for instance, Hacking claims that enumerations and counts somehow were always present in societies, even though at the beginning it was only to reach two main aims of governments: to impose taxes and military recruitment. Before the Napoleonic era, the greatest part of the data remained in a secret sphere and in the hands of administrators. Afterwards, large amounts of those data were printed and published. But the enthusiasm for numerical data was reflected in the censuses of the United States. The first US census made four questions for households; in 1880, in the tenth decennial census, there were 13,010 questions in various forms, addressed to particular persons, business firms, farms, hospitals, churches, etc. And this increase, albeit surprising, is minor compared to the increase of the printed figures. Behind this phenomenon were the new classification and enumeration techniques and the new bureaucracies with the authority and continuity required to implement the technology of counting. Somehow, many of the facts contemplated by the bureaucracies did not even exist. Kinds had to be invented so people could conveniently adjust to them and be counted and classified. The bureaucracy of statistics was imposed not only by creating administrative rules but by determining classifications within which people had to think about themselves and the actions that were available to them. Many of the categories with which we think about people and their behavior nowadays were put in that place by means of attempts to collect numerical data. The idea of recidivism—and hence the class of recidivist- for instance, appeared with the beginning of the quantitative study of crimes, around 1820. Likewise, during the nineteenth century, a canonical list of causes of death was established, which continues to this day with some modifications.

Criminals, prostitutes, the divorced, suicides, mentally ill, the insane, miserable, workers, autists, abused children, hyperactive children, female refugees, children TV viewers, the obese, are some of the kinds mentioned by Hacking as part of a long list that tabulates living beings—the *tableaux vivant* of which Foucault speaks—from the starting point of "*statistical information developed for purposes of social control*" (Hacking 1990: 6), among other things. Through the statistical study of populations, gigantic amounts of data are accumulated which can be effective to control or alter studies population in foreseen ways. The DSM-III, for instance, claims Hacking, is a diagnostic manual that presented classification schemes on the bases of which people were arrested according to the amount of complaints presented. There are two vectors here: on the one hand, the categories that can be used to count; on the other, the theoretical and practical reasoning of the experts when they face individual patients.

One of the examples that Hacking uses the most to show how a kind is molded in this way after a new classification is that of abused children. He has devoted to this example, as I have already mentioned, several articles and book chapters. The reasons for his choice are varied: in the first place, because it is a contemporary and pressing kind; secondly, because it is a relevant kind, in the sense that it has had great practical consequences in legislation, social work, the vigilance of the family, the lives of children and the ways in which children and adults represent their actions, their past, and that of their neighbors; thirdly, because in spite of its role in the rhetoric and the social policies of different tendencies, child abuse was initially presented as a scientific concept and it still pretends to remain so; finally, it is a profoundly moral kind, besides being scientific.

Child abuse as a way of describing and classifying actions and behaviors started with the debates and observations that took place in Denver, around 1960. It is a human kind that has been shaped and molded in the course of the past almost 60 years, during which it has been the center of intense debates, changing every few years.

Cruelty against children was one of the last great Victorian crusades, and it lost visibility around 1910, until it became invisible as a social problem. In the second half of the nineteenth century societies of aid and prevention were founded in cities such as New York, Liverpool and London. However, cruelty against children was never a radicalized issue. Many of the cases Victorians designated cruelty against children are included today within child abuse and vice versa, but both classifications are not identical. They are not so for several reasons. In the first place, there is a question of social class. Cruelty against children in the Victorian age was a matter of poor classes that caused damage to their children. Child abuse was presented, deliberately, as an issue without class distinctions. Secondly, Victorian activists, whereas they detested cruelty against children, were not scared by it. The opposite happens in the case of child abuse, which contaminates not only the child but society as a whole. Thirdly, cruelty against children was not a medical issue, but a question for the police, the courts of law and philanthropic societies. Child abuse was a medical issue from the start. The idea was proposed by doctors and the abusers were qualified as sick. This is a fundamental sense in which both classifications differ. Child abuse was viewed within a scheme of normality and pathology. Fourthly, sexual crimes or aggressions were not catalogued as cruelty against children, unlike child abuse (Hacking 1999).

The classification of child abuse became public for the first time at a meeting of the *American Medical Association* in 1961, on the part of experts, in this case pediatricians. A year later, the *Journal of the American Medical Association* published "The Battered-Child Syndrome", written by Henry Kempe and other pediatricians, and it was accompanied by a severe editorial about the long silence of society in the face of the reiterated appearance of lesions in small children, the objective proof of which was found in X-rays studies. In 1965, the *Index Medicus* added child abuse to its list of medical categories for its cataloguing. After 1968, in each state of the US there was a system to report abused children (Hacking 1991). 1971 could be the representative date of the public denunciation of the connection between child abuse and physical abuse, when Florence Rush spoke publicly about of this issue in the Radical Feminist Conference in New York. Likewise, 1975 could be singled out as the year in which sexual abuse within the family is recognized as a social problem. Child abuse, as a human kind was, as it can be seen, being molded and modified, changing radically and gradually encompassing a great variety of different actions and behaviors.

Many of these revelations were tremendously liberating for a group of people who had the chance, in the first place, to adjust and recognize themselves as members of kinds such as child abuse or child submitted to abuse, and to make public their degrading experiences. The kind of child abuse created a different world, a world where there is a new concept within whose terms to understand oneself. As Hacking points out, this might be the most powerful and provoking application of Nelson Goodman's idea of how classes make up worlds.

Child abuse was (and is) a new kind that changed the past of many people, their way of feeling who they are and how they came to be so. But it is also a clear example of the looping and feedback effect of a kind that is evolving. An example of interaction between experts and the classified, a sample of how the expert recognizes the classified person entirely and until the very end as an acting subject. It is in this sense that a feedback effect is produced, which in the case we are dealing with made it possible to move from the classification of cruelty against children to that of child abuse and later, to making the latter refer first to physical abuse but later sexual abuse and to continue to successively integrate other abuses, in a spiral of relations between those who classify and those who are classified and their behaviors, all of which keeps the notion under constant revision.

But recognizing the classified person entirely and to the end as an acting subject is not the only one of the two elements that are necessary for a power relation according to Foucault that we can identify in Hacking's proposal. Until now we have discussed this first element, which in some sense led us to a foray into the second, which we will develop in the next section. Let us recall that the second element necessary for a power relation is that a whole field of responses, reactions, results and possible interventions will open up in front of it. In other words, that it should face resistance.

5.3 Classifications, Looping Effect and Resistance

With regards to resistance to power, Foucault points out:

> [...] in power relations there is necessarily the possibility of resistance because if there were no possibility of resistance (of violent resistance, flight, deception, strategies capable of

reversing the situation) there would be no power relations at all (Foucault 1984a: 1539, 2000: 292).

His idea is that power relations necessarily provokes resistance, and that it does not come from the outside; it is contemporary and can be integrated to the strategies of power. The points of resistance are the reverse of power relations. But they are not a simple negative. Resistance is as inventive, as mobile and as productive as power.[1] People, unlike things, are organized and resist.

> [...] if there was no resistance, there would be no power relations. Because it would simply be a matter of obedience. [...] So resistance comes first, and resistance remains superior to the forces of the process; power relations are obliged to change with the resistance. So I think that *resistance* is the main word, *the key word*, in this dynamic (Foucault 1984b: 1559–1560, 2000: 167).

Resistance necessarily appears where there is power, because it is inseparable from the power relations in force. This entails the possibility of opening spaces of struggle anywhere, as well as enabling possibilities of transformation even within power itself. The power to affect and be affected are two aspects of any power relation. But there is something else, the resistance to power, the points of resistance that force and carry a mutation of the diagram.[2] These points of resistance play the role of adversaries, of targets, of ledges for apprehension, inscribed in power relations as its irreducible reverse. The power of resistance has the property of causing the diagram to change. The influx of power can always be modified, in certain conditions and according to a precise strategy.

Foucault analyzes the links between power relations and focal point of resistance in terms of tactic and strategy:[3] each movement of one of them serves as leverage for a counter-offensive of the other one. The link between both is not simply a dialectic schema. Resistance is not prior to the power it opposes. There is no logical

---

[1] *The Archaeology of Knowledge and the Discourse of Language* concludes a series of books that referred to knowledge, Foucault's first axis of thought. Foucault knew that this axis was not enough. Even though power was active in it, Foucault wasn't yet prepared to study it by himself. He still needed a long reflection. Six years later *Discipline and Punish* is published, and here he accesses the second axis, that of power. Year later, *The History of Sexuality. Volume I* contributes new elements: the relation between power and life; the specificity of resistance points. But even though in this text Foucault discovers these points, he still does not know what to do with them. This leads to an impasse in his work, the fact that he cannot find in his thought a place for the creation of the new, except as mutations of uncertain origin. The culmination of this impasse takes place in *The History of Sexuality. Volume II.* When he discovers subjectivation, he discovers a source of resistance points for an aperture of potentialities in a social field, and proposes a third axis, beyond knowledge and power, that will allow him to cross the latter's line.

[2] The diagram is the map, the cartography, always modifiable, always fluid, coextensive to the whole of the social field, that implies that, next to the points of affecting and being affected, there are other, relatively free ones, of mutation or resistance.

[3] There is a difference between tactic and strategy: tactics are the rationalities of power in particular cases; strategies are the grand systems or global models of power. Strategies are constructed on the basis of combinations and concatenations of local tactics. These comprehensive systems, or strategies, constitute institutional crystallizations starting from the interaction and combination of locally fluid power relations, and become terminal recognizable forms such as the State.

or chronological priority to resistance and, as I have said, it must present the same features as power: it must be inventive, mobile, productive. Like power, it is organized, solidified and cemented; it comes from below and it is strategically distributed. It can found new power relations, and new power relations can, inversely, promote the invention of new forms of resistance. Between power relations and struggle strategies there is reciprocity, indefinite linking and perpetual inversion. Is this feedback effect not the looping effect Hacking speaks of?

In this model, power is not conceived as totally negative, and struggles, on the contrary, conceived as attempts of liberation: not only is power, insofar as it produces truth effects, positive, but power relations are established everywhere because everywhere individuals are free. Struggles do not emerge fundamentally against power, but against some of its effects, against some states of domination in a space that, paradoxically, was opened by those very power relations.

In consonance with this idea, the looping effect is also partly constituted, according to Hacking, by a resistance—or a reaction, to put it in his words—to classification, which I understand as a resistance to the power of classification, even though he does not mention it explicitly. Those classified in a certain manner resist, passively or actively, the classification. And this is a key resistance, because if there is no reaction to the classification on the part of the classified the looping effect does not take place and we would be faced with what Hacking calls an indifferent kind. In this last case, there might be new classifications to substitute the older ones, but they would not be the product of the interaction between the classification and the classified. They would refer, in any case, to a model like that outlined for revolutionary (historical) nominalism, where there is a construction of new systems of classification according to certain interests of the experts in explaining the world, interests related to the problems that must be solved. There will be new categories, but the objects will not be historically constituted. The change in the systems of classification will have only one way, from the experts to the classified, not two, as is the case of dynamic nominalism, characteristic of the human sciences.

Hacking's looping effect is, in this point, a double example of what Foucault claims. In the first place, there are no power relations without resistances, which are more real and efficient when they are formed where power relations are exercised. Resistance to power does not need to come from outside to be real, but neither is it trapped because it is power's compatriot. It exists because it is right where power is: it is, like power, multiple and integrable in global strategies (Foucault 1980a). In the case of the looping effect, resistance is born where power is exercised, in the set of people to whom the classification affects. Not only the individual so classified, but also those he is related to, the matrix of institutions, the practices, etc., that surround him.

Secondly, Hacking's looping effect is an example of Foucault's claim that resistance is not only a negation: it is a process of creation; creating and recreating, transforming the situation, actively participating in the process (Foucault 1984b, 2000). The resistance is the creative side of power and is part of the idea, already mentioned, that power effects must no longer be always described in negative terms. Power produces, it produces reality, it produces environments and rituals of truth.

The individual and the knowledge that must be obtained from him correspond to this production.

Even though classified people tend in a certain sense to develop in the ways in which they are described, that is so say, in medicine, for example, patients tend to behave as experts expect them to, this is not always the case. Sometimes, says Hacking, they "*take matters into their own hands*" (1995a: 38). They resist.

An example of this comes hand in hand with the emergence of homosexuality and gay liberation. The word "homosexual", together with the medical and legal classification, emerged in the last half of the nineteenth century. For some time, the classification was property of the experts: physicians, psychiatrists and lawyers, who determined, at least superficially, what it meant to be a homosexual. But then the people so classified took charge and independently of what the medical-forensic experts attempted to do with their categories, the homosexual person became autonomous from the label. Being called or thinking about oneself as homosexual meant a great difference to many human beings in the past 150 years. People categorized as homosexuals took over the concept and changed the names, changed the meanings and changed the world. They took control of the kind and redefined it both in theory and in action (Hacking 2002: 345). Gay liberation was the natural result of this reaction which led to the labelled individuals to take control. One of the first characteristics of this liberation was gay pride and coming out of the closet. It became a moral imperative for people belonging to this class to identify themselves and attribute to themselves a chosen term. In this way, they also came into possession of knowledge about the phenomenon, beyond the fact that they were not the only people authorized to hold it, and they exercised what Hacking calls the autonomous behavior of the labelled person, who exercises pressure from the bottom up by creating a new reality that the expert must face.

It is not by chance, says Hacking, that this movement should find its vanguard in the United States, just like post-revolutionary-war France should have been the original site of human kinds. There are two reasons for this. In the first place, the US is a democratic society based on freedom of expression and information flow. Secondly, the important role played in this country by the social awareness of rights. Persons from a human kind demand their rights, or even those who are related to so-labelled individuals demand rights for the members of the kind (Hacking 1995b: 381).

This second reason allows me to introduce another one of Hacking's examples that we can mention regarding the resistance to classifications: autism.

The name "autism" was invented by Eugen Bleuler in 1908 to describe a family of symptoms characteristic of the group of schizophrenias. Later, the word was applied to some children previously considered mentally disabled, or even mute and deaf. This was a result of the research published by Leo Kanner in 1943, about a limited number of children. At the time, the mainstream conception influenced by the dominance of psychoanalysis in North American psychiatry was that the autistic child had a cold mother that could not express emotion towards the child. This doctrine has all but disappeared, and cognitive science came to play a fundamental role, insofar as autistic children have many deficits, including linguistic ones.

But what I want to emphasize here is that the case of autism—beyond its being controversial as an example of interactive kind—also shows, according to Hacking, how the changes in the family of the autistic child, intensely influence and one could even say damaged by the doctrine of the cold mother, contributed with their reaction to a rethinking of what at the time was understood as child autism, not because something new was discovered about the disease, but because the very behavior of those who surrounded the so-labelled persons reacted to the classification and the description. The looping effect present in autism has been driven mainly by activists' reaction or resistance to the classification. The people who have really influenced the shaping of our ideas about autism are personally connected with an autistic child or adult. They are relatives, profoundly concerned by what is happening to their children and they exercise resistance by creating something new.

Finally, the third example widely developed by Hacking: multiple personality disorder. In the first chapter of *Rewriting the Soul* Hacking remarks:

> As long ago as 1982 psychiatrists were talking about 'the multiple personality epidemic'. Yet those were early days. Multiple personality –whose 'essential feature is the existence within the individual of two or more distinct personalities, each of which is dominant at a particular time'- became an official diagnosis of the American Psychiatric Association only in 1980. Clinicians were still reporting occasional cases as they appeared in treatment. Soon the number of patients would become so overwhelming that only statistics could give an impression of the field (1995a: 8).

Ten years earlier, in 1972, multiple personality seemed a mere curiosity. Ten years later, in 1992, there were hundreds of people being treated for multiple personality disorder. Around 1970, strange behavior emerged, similar to the phenomena discussed a century earlier and for the most part forgotten. Multiple personality had decreased in the first half of the twentieth century, due to the fact that schizophrenia had become the dominant illness in the psychiatric-hospital system. But multiple personality returned with strength in the decades of 1980–90, mainly in the US. Some psychiatrists began to diagnose this disorder more frequently, and more people began to manifest these symptoms. The new wave of multiple personality differed substantially from previous editions. Both the diagnosed incidence of multiple personality in the population of the US as the amount of alters per patient proliferated. At the beginning they had the symptoms experts expected them to have. But later they became increasingly bizarre. At the beginning a person had two or three personalities at the most. A decade later, the number of alters had increased significantly and an average of 17 personalities were reported per patient.

This was incorporated into the diagnosis and became part of the standard set of symptoms. It became part of the therapy to obtain more and more alterations. Psychologists looked for the causes and created a pseudo-Freudian etiology of sexual abuse together with repressed memories. Knowing this was the cause, the patients recovered their memories. More than that: it became a way of being a person (Hacking 2007: 296).

Hacking explains the popularity of the syndrome on the basis of the growth of concerns about child abuse and sexual child abuse, as well as its particular use as defense in criminal cases. Multiple personality was absorbed by the movement of memory recovery that extended throughout the US in the 80s. Psychiatrists, patients, and other participants of this movement tended to interpret a series increasingly wider of adult disturbances as a consequence of repressed memories due to childhood trauma, combining memoro-politics[4] with sexual politics.

All the examples discussed until now show how on the basis of resistance, these power relations allow for a positive surplus of reality: or in more ontological terms, a production of being. The possibility of this production is linked to the total redefinition of power relations, i.e. what connects, on the one hand, the people that impose a power—sometimes the experts, sometimes the labelled individuals—and on the other, those subjected to that power and who eventually react against it—once again, sometimes the experts, other times the labelled individuals.

I underscore that sometimes the experts and other times the classified individuals because in the looping effect power circulates. Sometimes it is held by the experts who impose a classification, others by the labelled individuals, who react to the label and force the experts to adjust their descriptions to the new reality. Power is not, as Foucault claims, and I believe this is applied to this proposal of Hacking's, a general system of domination exercised by an element or group on another, and whose effects, thanks to successive derivations, would run through the social body as a whole. Power must not be looked for in the primary existence of a central point, in a single focus of sovereignty from which derived and descending forms would irradiate. It is the mobile pedestals of force relations which endlessly induce, due to their inequality, states of power—but they are always local and unstable.

Power relations are:

> [...] -strategic games between liberties-in which some try to control the conduct of others, who in turn try to avoid allowing their conduct to be controlled or try to control the conduct

[4] On the basis of the Foucauldian model of anatomo- and bio-politics, in *Rewriting the Soul* Hacking coins—to analyze the politization of memory—the notion of memoro-politics. Memoro-politics begins in post-revolutionary France with an alliance between an experimental-psychological science of memory and other secular forces of modernization. The program of localization of the cerebral function, marked by Broca's identification of the motor control of speech, was the last appearance of anatomo-politics. Experimental psychology may have started in laboratory psychology, as part of anatomo-knowledge, but with Ebbinghaus, as it became a statistical science, it no longer dealt with individual events or beings but with averages and deviations and it came to form part of the bio pole. Scientific psychology and psychiatry struggled with the ecclesiastical authority for the power to discipline and govern the soul, reclaiming from the Church the privilege of looking after it and disciplining it.

Hacking proposes to add to Foucault's two poles a third notion that can triangulate recent knowledge: memoro-politics. On this topic, he remarks: "*What is missing is pretty obvious. It is the mind, the psyche, the soul. Foucault spoke of "two poles of development linked together by a whole cluster of relations" [...] What I call memoro-politics is a third extreme point from which (to continue with the metaphor of mapping and surveying) we can triangulate recent knowledge*" (Hacking 1995a: 215).

"*Memoro-politics is above all a politics of the secret, of the forgotten event that can be turned, if only strange flashbacks, into something monumental. It is forgotten event that, when is brought to light, can be memorialized in a narrative of pain. We are concerned less with losing information than with hiding it. The background for memoro-politics is pathological forgetting*" (Ibid: 214).

of the others- and the states of domination that people ordinarily call "power" (Foucault 1984a: 1547, 2000: 299).

The field where power unfolds is not that of a lugubrious and stable domination: "*The struggle is everywhere [...] at every moment, we move from rebellion to domination, from domination to rebellion, and it is all this perpetual agitation that I would like to try to bring out*" (Foucault 1997: 184, 2003: 280). Failing to take this into account would be to neglect the relational character of power relations.

The analysis of power should be, then, says Hacking speaking about Foucault, "[...]*from the ground up, at the level of tiny local events where battles are unwittingly enacted by players who don't know what they are doing*" (1981: 74).

Those classified in a certain way -affected point- react to the classification, modify their behavior–resistance -and somehow exert a power—affecting point- which results in the need of a change in the classification so that it should adjust to the new reality of its members. Power acts as a chain process, it moves constantly. It is never focused here or there. Those who in one moment are affected, are in the following moment affecting, and this is what results in the looping effect. Power is exercised through a reticular organization, and people are always in a position to suffer or exercise that power, they are never the inert aim or consenting party of power, neither are they always the enforcing elements. In other words, power moves transversally, it does not lie still in individuals.

It is in this sense that Foucault conceives power in a capillary and omnipresent way,

> [...] not because it has the privilege of consolidating everything under its invincible unity, but because it is produced from one moment to the next, at every point, or rather in every relation from one point to another. Power is everywhere; not because it embraces everything, but because it comes from everywhere (Foucault 1976: 122, 1978: 93).

Foucault approaches power not as a substance or a process or a force, but as a *Lichtung*[5] which, as it opens a limited field of possibilities, governs the actions while still preserving its freedom (Dreyfus 1999: 89). The task is not to fix a primitive ontology, a definitively real stratum of historical reality, but to trace the mobile systems of relationship and synthesis that provide the conditions of possibility for the formation of certain orders and levels of objects and forms of knowledge about those objects. In this case, Foucault is even prepared to say that power does not exist (only the local exercise of power is real), claiming in this way that he is not offering a metaphysics of power. If power is taken in a nominalist sense—not as a real substance or a property, but simply as the name of a complex strategy or grid of intelligibility, this grid would be represented in different ways, and there is no final, privileged or foundational representation.

> One needs to be nominalistic, no doubt: power is not an institution, and not a structure; neither is it a certain strength we are endowed with; it is the name that one attributes to a complex strategical situation in a particular society. (Foucault 1976: 123, 1978: 93).

[5] *Lichtung* means divestment, a clearing in the forest, and it refers to the space open to the light that was preceded by darkness, the clear place that can be perceived in a thick forest.

Power is everywhere, and one of these places is in classifications. A place where the union of two elements much appreciated by Foucault, power and knowledge, takes place.

> Indeed, it is in discourse that power and knowledge are joined together. [...] We must make allowance for the complex and unstable process whereby discourse can be both an instrument and an effect of power, but also a hindrance, a stumbling-block, a point of resistance and a starting point for an opposing strategy. Discourse transmits and produces power; it reinforces it, but also undermines and exposes it, renders it fragile and makes it possible to thwart it (1976: 133, 1978: 101).

Nietzsche is particularly important on this point. A central hypothesis of the genealogical explanation of phenomena in terms of will power is that knowledge is power. Both for Nietzsche and for Foucault, the "is" that connects power and knowledge does not indicate that the relation of knowledge and power is one of predication, such that knowledge leads to power, but rather that knowledge is not obtained beforehand and independently from the use to which it will be applied to achieve power, that it is already a function of human interests and relations of power.

Foucault tries to make visible the constant articulation he perceives between power and knowledge. Traditional discourse does not admit the confluence of categories apparently as opposed as knowledge and power. Knowledge has to do with truth; power, with coaction. Knowledge belongs to the order of the necessary; power, of the contingent (Díaz 2003: 176). However, Foucault tries to show that it is not an opposition but an interaction. Between the techniques of knowledge and the strategies of power there does not exist any exteriority, even though each one of them has its specific role and they are articulated one on the other on the basis of their difference. It is necessary to end with the great western myth of the antinomy between knowledge and power, says Foucault in May of 1973. "*It is this myth that Nietzsche began to demolish by showing [...] that, behind constitutes and reconstitutes it: it is "a matrix within which elements of power come to function, are reactivated, break up*" (Foucault 2013: 279, 2015: 270–271).

Foucault is not satisfied by saying that power needs this or that discovery, this or that form of knowledge, but he claims that the exercise of power -and in the case of Hacking the exercise of power by means of classification, of categorization- "[...] *itself creates and causes to emerge new objects of knowledge and accumulates new bodies of information*" (Foucault 1980a: 51). Bodies of knowledge that, as Hacking admits in his article "The Archaeology of Foucault", "[...] *brings into being a new class of people or institutions that can exercise a new kind of power*" (1981: 73).

References

Díaz, E. (2003). *La filosofía de Michel Foucault*. (2ª. ed.) Buenos Aires: Biblos.

Dreyfus, H. L. (1999). Sobre el ordenamiento de las cosas. El Ser y el Poder en Heidegger y en Foucault. In E. Balbier, G. Deleuze, H. L. Dreyfus, M. Frank, & A. Glúksmann (Eds.), *Michel Foucault, filósofo* (pp. 87–103). Barcelona: Gedisa.

Foucault, M. (1975). *Surveiller et punir. Naissance de la prison*. Paris: Gallimard.

Foucault, M. (1976). *La volonté de savoir*. Paris: Gallimard.

Foucault, M. (1978). *The history of sexuality. Volume I: An introduction*. New York: Pantheon Books.

Foucault, M. (1980a). *Power/knowledge: Selected interviews & other writings 1972–1977*. New York: Pantheon Books.

Foucault, M. (1980b). *Language, counter-memory, practice. Selected essays and interviews*. New York: Cornell University.

Foucault, M. (1982). The subject and power. *Critical Inquiry, 8*(4), 777–795.

Foucault, M. (1984a). L'éthique du souci de soi comme pratique de la liberté. In M. Foucault (1994), *Dits et écrits* (Vols. 1–4, pp. 1527–1539). Paris: Gallimard, édités par D. Deferí & F. Ewald.

Foucault, M. (1984b). Michel Foucault, an interview: Sex, power and the politics of identity (« Michel Foucault, une interview: sexe, pouvoir et la politique de l'identité »). In M. Foucault (1994), *Dits et écrits* (Vols. 1–4, pp. 1544–1565). Paris: Gallimard, édités par D. Deferí & F. Ewald.

Foucault, M. (1995). *Discipline and punish. The birth of the prison*. New York: Vintage Books.

Foucault, M. (1997). *Il faut défendre la société. Cours au Collège de France, 1976*. Paris: Seuil-Gallimard.

Foucault, M. (2000). *Ethics, subjectivity and truth. Essential works of Foucault 1954–1984* (Vol. 1). New York: Penguin Books.

Foucault, M. (2003). *Society must be defended. Lectures at the Collège de France, 1975–1976*. New York: Picador.

Foucault, M. (2004). *Sécurité, territoire, population. Cours au Collège de France (1977–1978)*. Paris: Seuil-Gallimard.

Foucault, M. (2008). *The birth of biopolitics. Lectures at the Collège de France (1978–1979)*. New York: Palgrave Macmillan.

Foucault, M. (2009). *Security, territory, population. Lectures at the Collège de France (1977–1978)*. New York: Palgrave Macmillan.

Foucault, M. (2013). *La société punitive. Cours au Collège de France (1972–1973)*. Paris: Seuil-Gallimard.

Foucault, M. (2015). *The punitive society. Lectures at the Collège de France (1972–1978)*. New York: Palgrave Macmillan.

Hacking, I. (1981). The archaeology of Michel Foucault. In I. Hacking (2002), *Historical ontology* (pp. 73–86). London: Harvard University.

Hacking, I. (1984). Five parables. In I. Hacking (2002), *Historical ontology* (pp. 27–50). London: Harvard University.

Hacking, I. (1990). *The taming of chance*. Cambridge: Cambridge University.

Hacking, I. (1991). The making and molding of child abuse. *Critical Inquiry, 17*, 253–288.

Hacking, I. (1995a). *Rewriting the soul. Multiple personality and the sciences of memory*. Princeton: Princeton University.

Hacking, I. (1995b). The looping effects of human kinds. In D. Sperber, D. Premack, & A. J. Premack (Eds.), *Causal cognition: A multi-disciplinary debate* (pp. 351–383). New York: Oxford University.

Hacking, I. (1999). *The social construction of what?* Cambridge: Harvard University.

Hacking, I. (2002). How 'natural' are 'kinds' of sexual orientation? *Law and Philosophy, 21*(3), 335–347.

Hacking, I. (2007). Kinds of people: Moving targets. *Proceedings of the British Academy, 151*(p), 285–318.

Chapter 6
Experimentation and Scientific Realism: A Return to Francis Bacon

Scientific realism was all the vogue in 1980. I used the raging controversy about scientific realism as a peg on which to hang my plea for experiments.

Ian Hacking (2009: 147)

Abstract In this chapter *Experimentation and Scientific Realism. A Return to Francis Bacon*, mainly based on Representing and Intervening, I present Ian Hacking's critique of representation, of the theory and the realisms based on them. For Hacking, insofar as debates between realism and antirealism take place in terms of representation and theory, realism will not be capable of facing the challenges of antirealism. To correct this, he suggests shifting the discussion from realism to a different realm, experimentation, where—in his opinion—scientific realism is irresistible. I continue by presenting his defense of entity realism. The central notion in Hacking's argument for scientific realism is not reference but manipulation. In this sense, I analyze the two fundamental arguments on which Hacking bases his defense of scientific realism -intervention and coincidence—as well as his best-known examples, electrons and microscopes, respectively. Hacking's incursion in the realist debates of the 80s is a strategy to draw attention towards experimental scientific activity, to defend the vital and preeminent role of experimentation, and more precisely, the creation of phenomena in natural sciences.

Keywords Natural sciences · Experimentation · Scientific realism · Representing and intervening · Creation of phenomena · Ian Hacking · Michel Foucault

M. L. Martínez Rodríguez, *Texture in the Work of Ian Hacking*, Synthese Library 435, https://doi.org/10.1007/978-3-030-64785-8_6

6.1 Experimentation and Scientific Realism

The name of this node coincides with the title of both an article by Hacking (1982) and the last chapter of *Representing and Intervening*. It clearly identifies the true interest of the author on this subject: firstly, his treatment of experimentation, and in relation with experimentation, of scientific realism. Although Hacking's intention on this issue is frequently interpreted as a defense of scientific realism, he has insistently underscored in his latest work[1] that his incursion on this subject was due to the need to use it as a strategy to draw attention to his true objective: scientific experimentation, a subject that twentieth century philosophy of science had neglected and which at the time had begun to be timidly analyzed. In this sense, Hacking's purpose is to invert the traditional hierarchy of the theory over the experiment and show that it has its own life, independent from the first one.[2]

To this end, by means of the treatment and defense of a scientific realism of entities, he aims to highlight the importance of experimentation, of making in science, in opposition to the auxiliary role of representation and theory. To achieve this goal, he divides his book in two parts, one devoted to representing and the other to intervening, respectively.

In the first part, Hacking defends a realism of entities against a theoretical, representation-based realism, which he considers incapable of winning the battle against anti-realism. On this basis, he criticizes the exaggerated emphasis on theory and representation to the detriment of experimentation.

In the second part, he offers arguments to defend a realism of entities: intervention and coincidence that support his reflections on the role of experimentation in science. By means of his two most famous examples, the electron and microscopes, he illustrates the importance of making in scientific practice. This making allows for the creation of phenomena such as the laser, which do not exist until they are produced in the laboratory. Within this framework, Hacking approaches what we have identified as his most authentic interest, namely the analysis of the historical and situated conditions of possibility for the emergence of, in this case, laboratory-created phenomena.

Even when Hacking himself states that his defense and treatment of realism in 1983 is a way of drawing attention towards the subject he considers really interesting, I will follow here the order of his arguments and I will start by dealing with the subject of realism.

[1] See, for instance, Hacking (2007: 40, 2009: 147).

[2] Rouse (2002: 264) remarks that claiming that the experiment has a life of its own means at least: 1- Material practices of experimentation are not just means to observation but they always involve executions and skills that must be understood as having their own scientific significance; 2- Experimental work not only tests theories but, on the contrary, it responds to objectives, opportunities and restrictions that emerge within experimental practice; 3- The philosophical meaning of experimental practices and their results is not limited nor is it determined by theoretical interpretations; 4- Frequently, experimentation produces new and artificial phenomena, whose occurrence is not an exemplification of more general natural laws.

6.2 The Realism/Antirealism Debate

In 2009, Hacking claimed that rather than approaching the realism/antirealism debates in general terms, they should be discussed in the context of the styles of scientific reasoning, since those debates are produced by the introduction of new kinds of objects in specific styles of reasoning.

In 1983, however, Hacking talked about these debates in general terms and presented them as a house of cards based on the idea of knowledge as representation of reality. Atomism was the classical subject of debates about realism and antirealism. Antirealism about atoms was a physics issue. Until the end of the nineteenth century, statements about the ultimate structure of particles of matter were taken as intellectual tools, representational fictions, instead of real, existing entities that truly described the world. There was nothing questionable about it; since until then the minuscule structure of matter could not be proven; the only proof could be indirect: that the hypotheses might provide an explanation and contributed to making good predictions. In these circumstances, antirealism made sense, given the lack of knowledge or tools capable of testing the existence of such entities. Even though when Hacking (1983) writes his book the situation had changed – since those tools had become available—the antirealist ideal is still firmly rooted in the minds of some philosophers. Why? Because just like in the seventeenth century knowledge was conceived as a correct representation but one that couldn't be left in order to ascertain it corresponds to the world, many philosophers are still bound to a realism on the level of theory.

The best part of contemporary philosophical debates about scientific realism, claims Hacking, is posed in terms of theory, representation of reality, truth, explanation and prediction, but it says next to nothing about experiments, technology or the use of knowledge to change the world. He suspects that there cannot be any decisive argument in favor or against realism on the level of representation, on the semantic level. Consequently, he looks for an argument on another level of discussion, based on making, not saying.

When we move from representation to intervention, antirealism loses strength. The debates on the level of theories are necessarily inconclusive. Only on the level of experimental practice is scientific realism irresistible. Experimentalists only need to be realists about the entities used as tools. Hacking contrasts metaphysical questions related to scientific realism with those about rationality, that is, epistemological questions. The former lead to questions such as: are the entities postulated by theories of physics real? Epistemological questions deal with what we know, what we believe and what would be considered evidence. When arguing in favor of the realism of entities, Hacking focuses exclusively on metaphysical questions.

Incommensurability, transcendental nominalism, the substitutes of truth and styles of reasoning are part of the jargon of philosophers. They emerge when contemplating the connection between theory and the world. They all lead, according to Hacking, to an idealist blind alley, but none of them suggest a healthy sense of reality. All these arguments are conducted on the level of the discussion of theories. The problem with them is that they are trapped inside language.

Whence this obsession with representation, thought and theory to the detriment of intervention, action and experiment? What are the origins of these ideas of representation and reality? According to Hacking, human beings are representers. Not *homo faber*, but *homo depictor*. People produce representations. Everything that Hacking calls representation is public. When he talks about representations he means mainly physical objects: figurines, sculptures, paintings, engravings, objects produced precisely to be examined, contemplated. Representations are public and external, from the simplest sketch on a wall to the most elaborate theory about electromagnetic or gravitational forces. Theories, not individual statements, are representations.

The first proposition of Hacking's philosophical anthropology is that human beings are representers. Reality is an anthropocentric creation, a human creation, the second of human creations. The first one is representation. Once there is a practice to represent, there comes a concept of the second order, that of reality, a concept that has content only when there are representations of the first order. It can be argued that reality, or the world, were there before any representation or human language, but its conceptualization as reality is secondary. First there is representation, afterwards come the judgments about whether these representations are real or not, whether they are true or false. First there is the representation and then the real. First there is a representation and much later there comes a creation of concepts in terms of which we can describe this or that aspect where there is a similitude.

Alternative styles of representation pose the problem of realism. Different theories are different representations of the same reality. Each new theory attempts to assess a new realm of data. Mature science achieves a canonical set of approximations, the glue that keeps them together and that makes it possible to say whether the theory is true within that realm. The effort in the hard work of science is the construction of new hypotheses. This is a key to understand the philosophical interest in scientific realism. Departing from suggestions such as those of Thomas S. Kuhn and others, with the development of knowledge it is possible to inhabit different worlds. New theories are different representations, and therefore new kinds of reality. In the presence of alternative representations, there emerges the problem of thinking about what is real; then science has to elaborate criteria about what counts as a correct representation of reality.

It is important to clarify what Hacking is talking about when he discusses realism. Many realist scientists and philosophers suppose that the ultimate aim or ideal of science is a single theory about the world and argue in favor of a realism in terms of convergence towards this grand truth, or at least of convergence towards some value called "truth". Hacking does not subscribe to this idea; on the contrary, he believes that there is nothing that requires a single and richer theory about what the world is like. The universe is too complex to be described by a single theory. However, it is possible to imagine an infinite formation of descriptions of the world, independent from each other and at the same true with regards to it.

Some philosophers of science have chosen to narrow down the subject of realism and speak of scientific realism. A certain kind of scientific realism about theories posits that they are true or false, or candidates to truth or which aspire to truth and sometimes approach it. The objective of science is the discovery of the internal constitution of things and the knowledge of what inhabits the most distant confines of the universe. In this sense, some anti-realists claim that theories should not be taken literally, that they are intellectual tools and that they are at the most legitimate, adequate, useful, good instruments for work and prediction. Others argue that theories must be accepted literally, even if we have no reason to believe they are correct.

Another version of scientific realism posits that theoretical entities really exist. It is called entity realism, and Hacking includes among these entities processes, states, waves, currents, interactions, fields, black holes, affirming in this way the existence of at least some of the entities that make up physicists' stock. This realism posits that several theoretical entities really exist, which means that there is a real ontological level to which they belong, even though it is possible to have several competing or even contradictory theories about them. He also says that the existential statements of a theory must be considered true, and also that such entities would exist even if there was no theory that referred to them, nor, therefore, any knowledge about them. Such entities are independent from the theory that postulates them. The reasons to believe in these statements come not only from the scientific theories themselves, but also from a philosophical argumentation about the possibility of observing in experimentation the same kind of entities on the basis of different physical processes. Protons, photons, force fields and black holes are as real as turbines, whirlpools and volcanoes, according to Hacking. The corresponding antirealism denies that such entities exist and affirms that they are fictions, logical constructions or parts of an intellectual instrument to think about the world. According to this, theories are constructed about states, processes and minute entities with the single aim of having the capacity to predict and produce events that interest us.

There is no agreement about the precise definitions of both kinds of realism discussed here. Both have an implied gnoseological component: sometimes there are good reasons to think that our theories are true, or that some unobservable entities, such as electrons, exist. Both could seem identical. If someone believes in one theory, do they not believe in the existence of the entities about which they speak? If they believe in some entities, shouldn't they be described in some accepted theoretical sense? This apparent identity is illusory. Both kinds of realism can be complementary, but they are distinct. It is possible to be a realist about theories but an antirealist about entities. Likewise, it is possible to think that many theoretical entities exist, but that theories about them do not need to be true. In fact, like Hacking, many experimental physicists are realists about the entities but not about the theories.

This is to say, there is an important experimental contrast between entity realism and theory realism. Very few experimenters would deny what the latter states: that science looks for true theories. However, the search for truth is about an indefinite future. To try to argue in its favor is to lock oneself up inside the world of representations, and it is not surprising that scientific antirealism will always be lurching behind. Hence Hacking's insistence that the problem of realism has not been adequately posed. That if it is presented only in terms of theories and their capacity or function of representation, at most it might be possible to establish a realist position with respect to theories. In *Representing and Intervening* he claims that there can be no decisive argument in favor or against realism on the level of representation. *"Realism and anti-realism scurry about, trying to latch on to something in the nature of representation that will vanquish the other. There is nothing there*" (145).

However, experimental science could actually lead to a realism about the entities postulated by the theories. Electrons are polarized to investigate the weight of the neutral current. What results from this—Hacking claims—is a firm realism about electrons which would be hard to attack, even by a skeptic or antirealist about the exact truth of theories regarding the entities being manipulated. His experimental argument in favor of scientific realism about entities trusts that entities are regularly used to achieve effects or to study other phenomena: "[…] *that reality has more to do with what we do in the world that with what we think about it*" (Ibid: 17), because "[…] *realism is more a matter of intervention in the world, than of representing it in words and thought* […]" (Ibid: 62).

The way in which experimenters are scientific realists about entities is completely different from the ways in which they can be realists about theories.

6.3 Entity Realism

After his criticism of semantic realism,[3] and aiming to underscore the importance of experimentation in science, Hacking develops an experimental argument to defend non-observable entity realism. He claims that certain entities can be characterized using low-level generalizations about their causal properties and the senses in which they interact with other parts of nature. Such generalizations do not constitute anything like what is understood as theory; rather, they are a set of shared beliefs about the entity, a set of beliefs used together with a theory of reference of the kind defended by Hilary Putnam. These home truths can be expressed in different theories and incompatible models that can be used simultaneously in different parts of the same experiment, and there is no reason to think that their intersection would constitute a theory, or to think that it is necessary to be a realist about this theoretical background or model on which these home truths are based.[4] Moreover, since these

[3] See Martínez (2009).

[4] Margaret Morrison (1990) notices some difficulties that emerge from this characterization. They have to do with the problem of separating these home truths from their theoretical background and establishing their epistemic authority as generalizations to be interpreted in a realist sense. She claims that it is not clear that such a group of supposedly neutral theoretically generalizations could be isolated – as Hacking claims—and that even if this were possible the question would

models are frequently inconsistent, they can be considered to have instrumental value in some contexts but not in others. For example, first it is conjectured that there is an entity of a certain kind, and later instruments to see it are developed. Shouldn't even positivists accept this evidence? No. *The positivist says that only theory makes us suppose that what the lens teaches rings true. The reality in which we believe is only a photograph of what came out of the microscope, not any credible real tiny thing* (Hacking 1983: 186).

On the contrary, the great majority of experimental physicists are realists regarding some theoretical entities even though they do not need to be so. "*Experimenting on an entity does not commit you to believing that it exists. Only manipulating an entity, in order to experiment on something else, need do that*" (Ibid: 263).

Clearly, it is not the notion of reference that shapes the central argument for Hacking's realism but rather the manipulation of entities in scientific practice. Even though Hacking admits that nature is not constituted by human manipulability, he thinks it provides the strongest evidence for the reality of an unobservable theoretical entity, and for this reason he occasionally contrasts this kind of cases with others which merely involve the construction of models of phenomena. Since scientists tolerate the existence of multiple models of the same phenomenon, even if they are internally inconsistent, it is difficult to argue that these models are realistically interpreted.

That experimenters might be realist about entities does not mean they are right. Instruments that rely on the properties of electrons to produce precision effects can be made in several different ways. That is to say, its reality cannot be inferred from the success obtained with the electrons. Instruments are not built to later infer the reality of electrons, as when a hypothesis is tested, and then people believe in it because it passed the test. This orients the temporal order in the wrong direction. Firstly, according to Hacking, a device is designed on the basis of a small number of truths about electrons, in order to produce another phenomenon to be researched (Ibid: 265).

People do not believe in electrons because it is possible to predict how devices work. The belief about the reality of electrons emerges when regularly – and for the most part successfully— new kinds of devices are constructed, using diverse, well-understood causal properties of electrons that make it possible to interfere in other, more hypothetical areas of nature. (Ibid).

---

persist of whether they offer sufficient theoretical grounds to completely characterize the results he describes as intervention. Whereas Hacking is right in stating that interference and intervention are the substance of reality, according to Morrison the traditional problem of scientific realism still stands in terms of whether it is possible to correctly identify and to attribute properties to said substance and what kinds of epistemic claims can be made. Hacking solves these problems of theoretical instability by means of home truths and a Putnamian theory of reference, but Morrison claims that a small group of truths cannot yield a substantive characterization of entities and that they are not complex enough to allow the kind of successful engineering that Hacking claims. On the other hand, if the group of theories is extended to include causal interactions and processes based on a higher level of the theory, instability emerges with the subsequent problems of theoretical realism.

Only then do electrons lose their hypothetical category; when they are used to investigate something else, when it begins to be possible to do things with the theoretical entity.[5] To begin with it could be measured, and later something can be sprayed with it. There will be a series of incompatible explanations, all of which, however, will coincide in the description of the causal powers that could be used to intervene with said entity in nature.

Physics is Hacking's science of preference to illustrate his conviction about entity realism and, as we see, his favorite example is the electron. According to Hacking, electrons made him a realist, because he could spray with them a niobium ball to alter its charge. *"From that day forth I've been a scientific realist. So far as I'm concerned, if you can spray them then they are real*" (Ibid: 23).

The direct proof is that they can be manipulated utilizing low-level causal properties. Electrons exist even though one is not capable of providing true descriptions of them beyond a purely phenomenological level. During the first stages of the discovery of an entity such as this one, it is possible to test the hypothesis that it exists, it is possible to doubt its existence. This is what happened even after Thomson measured the mass of electron corpuscles and Millikan their charge. It was important to be sure that both were measuring the same entity. Greater theoretical elaboration was needed. Later, the best reason to think about its existence was its success in explanation; postulating its existence could explain a great variety of phenomena. Finally, success was not based on explanation but rather on the fact that as its causal powers became better understood, new devices were built that achieved well-understood effects in other parts of nature. When it becomes possible to use them to manipulate these other parts systematically, the electron ceases to be a hypothetical, theoretical entity and it becomes experimental (Ibid: 262).

Electrons are not useful to organize thoughts or save observed phenomena, but they can be used to create new phenomena. They are tools manipulated in the engineering of the scientific experiment destined to extend the frontiers of knowledge.[6]

There is a family of causal properties (mass, charge, spin) in terms of which experimenters describe and use electrons to investigate something else. When such properties are used in the exploration of physical reality, it is possible to see the engineering of scientific instruments at the service of science. Such properties are also relatively resistant to theory changes.

---

[5] Hacking uses the term theoretical entity to refer to entities postulated by theories and which cannot be observed.

[6] On this issue Dudley Shapere (1982) remarks that physicists talk about observing even when they use devices where the senses cannot play an essential role. He mentions as an example the attempt to observe the inner part of the sun by means of neutrinos emitted during the process of fusion, and where what counts as observation depends on the theory used. Much more would be known about the sun if more was known about the solar neutrinos that reach the earth. Physicists gather the sun's neutrinos in a huge abandoned mine that was filled with a cleaning fluid that captures neutrinos. A few of them would form a new radioactive nucleus. Even though in this study the extension of the manipulation of neutrinos is far below the manipulation of the electron in Hacking's examples, neutrinos are clearly being used to investigate something different from themselves.

This engineering is the best proof of scientific entity realism because, according to Hacking, what matters is not so much to understand the world but to change it. This is the first argument he offers to defend realism: intervention.

It is necessary, however, to clarify some points about this experimental argumentation, since certain ambiguities have led, occasionally and according to Hacking himself, to a misinterpretation. In the first place, as we have already pointed out, Hacking does not claim that the experimental argument is the only evidence of the reality of an entity. It is the strongest argument, the one that can be the most convincing or irresistible on the basis that scientists not only feel that the entities they use are as real as their own hands, but also because they can do something with them. But it is not the only argument. It is the claim that, when said entity is regularly used to transform another aspect of the universe, antirealism about this entity does not make sense, or at least does not make so much sense. His experimental argument for entity realism can imply a sufficient condition to claim that an entity exists. But it does not imply a necessary condition. There can be many kinds of evidence of the existence of an entity. What Hacking is affirming is that the ability to manipulate an entity is the best evidence. This leads to a second question. His experimental argument is not a statement about the reality of theoretical entities in general, but about the reality of this or that particular entity. Hacking is a philosopher of particular cases. Thirdly, that an entity cannot yet be manipulated as a tool to investigate something else means that there is to that moment an irresistible argument for its existence. It does not mean, however, that in those circumstances there is no such argument, or that it is not possible to think reasonably about the existence of such an entity. The experimental argument does not imply that if there is a lack of manipulability the entity does not exist; it means that we do not have an irresistible proof of its existence. Fourthly, he is not talking about mere manipulation. He is talking about entities that are regularly manipulated to produce a new phenomenon and to investigate other aspects of nature. The manipulation Hacking speaks about has a purpose, or rather, several purposes: create phenomena, investigate other aspects of nature, interfere with nature by creating effects in other parts of it. It must be borne in mind that Hacking speaks about "regularly" manipulating the entity in question.

6.4 Creation of Phenomena

The eyes and the hands are not too far apart, says Hacking. Dewey[7] and Rorty rightly point out that the metaphor of vision carries ideas of correspondence, representation and the mirror of nature. It must be borne in mind what Dewey calls the

[7] In *Representing and Intervening* (1983: 62), Hacking comments that his remark that realism is a question of intervening in the world rather than representing it in words and thought owes much to Dewey, who despised dualisms such as mind/matter, theory/practice, thought/action, fact/value, and mocked what he called the spectator theory of knowledge. He considered it the result of a

spectator theory of knowledge, which he criticizes to undermine the conception of knowledge and reality as a question of thought and representation.[8] Science is not a spectator sport. It is a game to be played, and those who play football, tennis, basketball, etc., do not infer the existence of the ball: they kick it, move it, throw it.

To experiment is to create, to produce, to refine and stabilize phenomena.[9] It is to produce phenomena that do not exist naturally in a pure state or that do not exist until they are constructed. Phenomena that are the touchstone of physics, the keys to nature and the source of much modern technology. It is creating and not simply discovering phenomena, because these are difficult to produce in a stable manner. There are a series of different tasks. It is possible to talk about designing an experiment that might work, of making the experiment work. But perhaps the hardest thing of all is to tell when an experiment works; this is why observation, in the usual sense in Philosophy of Science, plays a relatively minor role in experimental science. Here what matters is not annotation and information of the readings of instruments,[10] it is not to describe or to inform, but the ability to distinguish that which is incorrect, anomalous, instructive or distorted in the experimental equipment. The experimental scientist is not the passive observer depicted in traditional philosophy of science, but an alert and perspicacious person. Observation and experiment are not the same thing, they are not even the poles of a continuum. To experiment is to interfere in the course of nature. Interference and interaction are the substance of reality.

Experimentation has a life of its own which interacts with speculation, calculation, model construction, invention and technology in different ways. However, whereas the speculator, the calculator and the model constructor can be antirealists, the experimenter, according to Hacking, must be a realist.

It is necessary to make, not to just look, to learn to see, for instance, through a microscope. Practice – making—develops the skill to distinguish between visible artefacts of the preparation or the instrument, and the actual structure seen through the microscope. It is this practical skill which engenders conviction (Hacking 1983: 191).

Philosophers frequently imagine that microscopes lead to a conviction about something because they help to see better, but this is only part of the story. What matters is not this, but what can be done with a specimen under the microscope and what can be seen by making it. One learns to see through the microscope by making

---

wealthy class that thinks and writes about philosophy, in opposition to a class of entrepreneurs and workers that not only see but also manipulate.

[8] Criticism that, according to Hacking it did not have the expected success since "*He should have turned the minds of philosophers to experimental science, but instead his new followers praise talk*" (1983: 63).

[9] In *Scientific Reason* (2009:111), Hacking remarks that having spoken about the creation of phenomena seems an exaggeration, and that it might be better to say that phenomena are purified or achieved.

[10] An activity emphasized, among others, by Bruno Latour and Steve Woolgar (1979).

something, not just looking. The new ways of seeing involve learning through making, not just through a passive gaze.

Microscopes lead to conviction thanks to the interactions and interferences that they make possible. It is the power to use unobservable entities, to make something with them, what makes us believe they are there. Different instruments (for instance, microscopes[11]) which use very different physical principles lead to observing the same structures in the same specimen, which allows us to conclude that they are independent from the theories that allow their observation and identification. The role of the theory used to build the instruments that allow the phenomenon to be seen is relatively minor. In general, philosophers tend to think that nothing which cannot be touched or seen can be a theoretical or inferred entity. On the contrary, physicists talk about observing the actual entities that philosophers say are not observable.

This argument, that of coincidence – the second Hacking uses to support his defense of realism—states therefore that:

> It would be a preposterous coincidence if, time and again, two completely different physical processes produced identical visual configurations which were, however, artefacts of the physical processes rather than real structures in the cell (1983: 201).

If the same features of a structure can be seen using different physical systems, then there are very good reasons to say that this is real,[12] instead of saying that this is an artefact.[13] If these techniques relate processes considered independent, unrelated, it would be strongly improbable that they would produce the same visual configurations. But Hacking does not propose this argument as the only basis for the conviction about what is seen through the microscope. This is a convincing visual element that combines with other intellectual modes and other kinds of experimental work. Hacking mentions some others: (1) success in systematically removing aberrations and artefacts, (2) the possibility of interfering with the structure that appears to be visible, in purely physical senses and (3) a clear understanding of the greatest part of physics used to build the instruments that allow visualization – even though this theoretical conviction plays a relatively small role. These and other interconnected low-level generalizations make it possible to control and create phenomena with the microscope.

---

[11] In their different types, such as optical, of interference, of polarization, of phase contrast, of direct transmission, of fluorescence, etc.

[12] Hacking explains, following Austin, that the real word can change content depending on the context and the phrase where it belongs. In the case of this argument, real is considered in opposition merely to an artefact of the physical system, whereas in the case of the first argument on which his defense of realism is based – intervention—it is used as opposed merely to a tool of thought.

[13] A traditional answer to the problem of knowing what it is that is seen and what is real as opposed to an artefact would seem to require a substantial trust in theory, something to which Hacking is reticent. He tries to overcome this difficulty remarking that, even though some theory of light is essential to build microscopes, it is usually a low-level theory, where engineering is the most important component. In the cases of real observation he minimizes the role of theory, claiming that visual exposition is resistant in the face of theory changes

6.5 "Saving Phenomena"?

"Saving the phenomenon" is an expression translated from Greek that literally means: to reconcile the observed phenomenon and a theory that contradicts it. Its reference is astronomy. The phenomenon does not confirm the movements calculated by celestial models, and both are regularly reconciled by the addition of epicycles to the latter. During the seventeenth and eighteen centuries, the expression was adapted to other sciences, but with a verb change. Playing with the term *salve* in Latin, it refers to solving the phenomenon. "To solve" retains its old sense of solving a problem in geometry, and therefore to solve the phenomenon is to construct an empirically adequate theory, to give an explanation of the phenomenon. At the beginning of the twentieth century, with Pierre Duhem's book *To save the phenomena* (1908), and later with a chapter of the same title by Bas van Fraassen (1980), this theme is reanimated. According to Hacking (1989: 577), in this work, van Fraassen claims that all science aims only to save phenomena, to an empirical adjustment, to the ability to derive observable phenomena.[14] Hacking does not believe that natural science aims at this but rather to the manipulation of and interference in the world in order to understand it and change it. To save the phenomenon appears then as a merely subsidiary aspect of scientific activity. There are perhaps only two branches of science where the label saving the phenomenon has a central place: astronomy and astrophysics. This is, according to Hacking, because in these sciences, technology has changed radically since antiquity, while their method has remained exactly the same: to observe celestial bodies, to construct models of the macrocosm, to try to make observations and models agree. They continue to be the perfect paradigm in the Style 3 of Alistair Crombie.[15] In contrast, the methods of the natural sciences have suffered a profound transformation, mainly in the seventeenth century. The transition in these sciences moves precisely towards the experimental method, the interference in nature, the creation of new phenomena. According to Hacking, van Fraassen is wrong in claiming that all science is a question of empirical adjustment and saving phenomena. And this error is due to the fact that he—like many other philosophers—is completely theory-oriented, and blind to the experiment.[16] The natural sciences are not a question of saving but of creating phenomena. But in astrophysics they cannot be created, they can only be saved, according to Hacking. When the reality of the entities that interfere in other aspects of nature and investigating the interior constitution of matter is posited, when they are regularly used as tools, as research instruments, then they are being considered real. This

[14] As my anonymous reviewer points out, van Fraassen (2008, among other works) also vindicates the importance of pragmatic aspects in his proposal.

[15] Let us bear in mind that Style 3 in Crombie's list is the hypothetical construction of analogical models.

[16] In later works, van Fraassen revised his position on these issues.

cannot be done with the objects of astronomy and astrophysics,[17] and if it is done, mistakes can be made.

> I must now confess a certain scepticism, about, say, black holes. I suspect there might be another representation of the universe, equally consistent with phenomena, in which black holes are precluded. I inherit from Leibniz a certain distaste for occult powers. Recall how he inveighed against Newtonian gravity as occult. It took two centuries to show he was right. Newton's aether was also excellently occult. It taught us lots. Maxwell did his electromagnetic waves in aether and Hertz confirmed the aether by demonstrating the existence of radio waves. Michelson figured out a way to interact with the aether. He thought his experiment confirmed Stokes's aether drag theory, but in the end it was one of many things that made aether give up the ghost. The sceptic like myself has a slender induction. Long-lived theoretical entities, which don't end up being manipulated, commonly turn out to have been wonderful mistakes (Hacking 1983: 275).

It is not the case that the experimental argument is the only viable argument for scientific realism about non-observable entities, neither that it is conclusive; but it is the most, and perhaps the only, irresistible one. Unfortunately, Hacking does not explain why it is the only irresistible one, nor does he analyze other kinds of argument which, albeit less irresistible, could be taken as supporting realism in particular cases.

On the basis of his experimental proposal, Hacking argues explicitly in favor of antirealism – of the antirealism that corresponds to his particular realist conception—in astrophysics, because even though it is already possible to experiment on the moon and nearby planets, it is not possible to experiment in or with the sun. According to him, galactic experimentation is science fiction, whereas extra-galactic experimentation is a bad joke (Hacking 1989: 559).

His example deals with gravitational lenses. Hacking questions, in his article of 1989, the existence of such lenses. He says that the optimism of their proponents disappears in a pile of inconclusive data, failed predictions and incompatible models. But this theoretical fragility is but a symptom of something deeper. He is skeptical about gravitational lenses because they are incapable of ontological promotion by means of experimentation.[18] In astrophysics there are only models, in every

[17] Dudley Shapere (1993) has criticized Hacking's position on these sciences, arguing – among other things—that he has been very selective in his analysis, omitting visions and information which are directly relevant to the case and which undermine his argument. Hacking's problem, according to Shapere, even though it is deeper and goes beyond the case of astronomy, is that he has misunderstood the role of experiment in science.

[18] Dudley Shapere – in the aforementioned article of 1993—attacks Hacking's argument saying that even though it is correct in that these lenses present difficulties, these are not reasons to abandon trust in the possibility of obtaining realist knowledge in astronomy. On the other hand, he accuses Hacking of forgetting that research with gravitational lenses is a new area suffering from the difficulties and doubts of the fields on the borders of sophisticated science, not only in astronomy but in science as a whole.

Shapere is not alone in attacking this argument of Hacking's. The astrophysicist Edwin Turner reports in 1988 on the study of gravitational lenses as a matter of fact, a complete history of engineering details. Turner argues that lenses could act like natural telescopes of cosmic dimensions that would allow astronomer to directly determine the dimension and age of the universe. In this

possible stratum of investigation. There are no propositions of detail that could be more than models. They are not literally true. Neither are they truth-convergent, since all that can be had is more models. For this reason, he is antirealist about astrophysics (Ibid: 576).

In the light of what Hacking has underscored on this point in *Scientific Reason* (2009: 154), his declaration of antirealism must be understood in the following sense: his experimental argument in favor of realism cannot be invoked in the case of gravitational lenses. This does not mean to affirm offhand that they are definitely not real, but rather to vindicate that there are more irresistible reasons to affirm the existence of polarized electrons than of gravitational lenses.

It is important to notice that Hacking uses theories, models or phenomena as a possible framework within which to think about science. But he claims that one must be an antirealist about them because it is possible to work with inconsistent models – if we take them in the sense of making literal claims about the world—in order to solve the same class of problems. They are intellectual tools that help us understand phenomena and build parts and pieces of experimental technology. They allow us to intervene in processes and to create new phenomena never imagined before. But what makes things work are the entities that are producing these effects. The entities are real, they produce the effects. The model can help organize the phenomena in our minds, but it is not a literal representation of how things are in reality.

Hacking severely criticizes astrophysicists' practical reports because of the way they create certainties by means of verbal manipulations that are not accompanied by additional evidence. All these defects in scientific argumentation are due to a series of scientific practices that cannot establish the real. He suspects there are no branches of natural science where the presence of models is more endemic than in astrophysics, or one in which at all levels they are so central.[19]

way, they would become a new tool to tackle fundamental problems. Thus, lenses would pass Hacking's test, becoming tools, just like electrons. In my view, this is not the experimentation Hacking discusses, which has effects on another thing and is endowed with a certain regularity.

[19] Shapere (1993) criticizes Hacking once more, now about the issue of models. Firstly, because he understands that these also have a role in the conception and interpretation of experiments which is central to the version of realism that Hacking defends. Secondly, because the fact that models occur in astrophysics and in the field of gravitational lenses does not exclude the possibility of offering a realist account of the observed phenomena. Finally, because astronomers are not satisfied by the use of multiple and eventually inconsistent models, but try to reach a single, more adequate one. Ultimately, says Shapere, this should not lead to antirealism but to agnosticism; it is possible to have reasons to believe that certain objects—such as gravitational lenses—exist, even though certain facts might prevent us from knowing at least some things about them.

With regards to the first criticism, I do not believe Hacking does not share the idea that models are used in other areas; in fact, he mentions examples of other fields such as genetic engineering, where they are used. But what Hacking underscores is that its use is endemic in astronomy and that in this science it would be impossible to advance beyond the model because it would be impossible to intervene. This is also true with regards to Shapere's second criticism: it is not working with models that excludes realism according to Hacking, but the fact that it is impossible to do more than that. With regards to the third, the issue is not that astronomers are not satisfied and that they

6.6 Representing and Intervening in the Natural and Human Sciences

In order to conclude this chapter, I think it is necessary to reflect on the role that theory and representation play in Hacking's proposal on the one hand, and their relationship with intervention on the other. My impression is that, in the first place, it is not quite clear that theory is not needed in his argument on experimentation in the natural sciences, and secondly that these questions become even more pressing when we analyze the role it plays in his proposal for the human sciences.

With regards to the first issue, the question emerges of whether his proposal does not use more theory than he is ready to admit on several occasions. In the case of the microscope, for instance, despite Hacking's claims that it is possible to tell apart the theory of the device and the theory of electrons, it is not evident that this distinction can be made in a radical manner within the context of practice of a research. This is the case of laboratory sciences in general, where Hacking thinks that even though experimental material, that is to say, devices, instruments, substances and objects to be investigated, flanked on one side by ideas and on the other by symbols and manipulations of symbols, is more central than what certain specialists show in social studies of science, this is not enough reason to lose sight of the relation theory-experiment. Laboratory sciences are of necessity theoretical. Laboratory science includes the whole theoretical superstructure and the intellectual realizations that finally answer to what happens in that institution.

With regards to the second issue, in *Representing and Intervening* Hacking emphasizes the role of intervention in scientific activity and assigns to theory a secondary role. In his works about human sciences, Hacking claims that intervention is the substance of reality, but – in my view—it is understood in a different way, mainly because its relationship with theory changes. In his work on transient illnesses, we saw how theory, representation, conceptualization, take on a more important role; they are no longer simply a set of low-level generalizations or home truths. Here Hacking suggests that the discourse, the category, contribute to the emergence of new realities. It is not that suicides, homosexuals, abused children, do not exist, but until they are classified as such they are not suicides, homosexuals, or abused children, respectively. It is not that these real facts do not exist before they are labelled, but rather that they are not such facts. There are actions, but not a kind of person such as a homosexual, suicide or abused. They are kinds of people that end up being reified. Discourse intervenes both in the emergence of this new reality and in the alterations it suffers due to its interaction with the ways in which it is classified. Certainly, what impinges on these persons is not merely the theories about them, but the institutions and the practices to which they are subjected; but do they not result from conceptualization, from the sustained theory? Hacking himself

look for a single model, because Hacking does not defend the existence of a single theory, but rather that the question is if insofar as it is not possible to work but with models, it would ever be possible to solve this inconsistency.

claims that the distinction between word and thing is fuzzy in the human sciences, that the possibilities of their objects (human action) deliberately depend on the possibilities of description. However, he claims in these same writings that discourse does not do this work. Even though it may not do it, it plays a role that it does not seem to have here, in the natural sciences. And what about intervention? The evidence of the reality of the electron is that we can use it as a tool to intervene in other aspects of the universe. In the human sciences, according to Hacking, questions also seem bound to intervention and causation, but what is it that intervenes? The practices and the institutions which are themselves related to the classification? And, what reality is this intervention proving? That of which intervenes – as in the case of the electron—or that which results from that intervention – as in the case of the transient illness or child abuse?

References

Hacking, I. (1982). Experimentation and scientific realism. In R. Boyd, P. Gasper, & J. D. Trout (Eds.) (1991), *The philosophy of science*. Cambridge: Blackwell, p. 247-260.

Hacking, I. (1983). *Representing and intervening*. Cambridge: Cambridge University.

Hacking, I. (1989). Extragalactic reality: The case of gravitational lensing. *Philosophy of Science, 56*, 557–581.

Hacking, I. (2007). On not being a pragmatist: Eight reason and a cause. In C. Misak (Ed.), *New pragmatist* (pp. 32–49). Oxford: Claredon.

Hacking, I. (2009). *Scientific reason*. Taipei: National Taiwan University.

Latour, B. & Woolgar, S. (1979). *Laboratory life: The construction of scientific facts*. Beverly Hills: Sage Publications.

Martínez, M. L. (2009). *Realismo científico y verdad como correspondencia; estado de la cuestión*. Montevideo: Facultad Humanidades y Ciencias de la Educación.

Morrison, M. (1990). Theory, intervention and realism. *Synthese, 82*, 1–22.

Rouse, J. (2002). *How scientific practices matter. Reclaiming philosophical naturalism*. Chicago: University of Chicago.

Shapere, D. (1982). The concept of observation in science and philosophy. *Philosophy of Science, 49*(4), 485–525.

Shapere, D. (1993). Astronomy and antirealism. *Philosophy of Science, 60*, 134–150.

van Fraassen, B. (1980). *The scientific image*. Oxford: Oxford University.

van Fraassen, B. (2008). *Scientific representation: Paradoxes of perspective*. Oxford: Claredon.

Chapter 7
On Foucault's Shoulders

[...] c'est Foucault qui m'a poussé dans ma nouvelle voie.

Ian Hacking (2002b: 13)

[...] I feel no inconsistency between my analytic instincts and my ability to use some aspects of Foucault.

Ian Hacking (1990: 71)

Abstract Chapter *On Foucault's Shoulders*, starts by proposing that the subject that awarded Ian Hacking more visibility and consideration, dealt with in Representing and Intervening, is not the one that he has been most interested in nor the one to which he has devoted more time and publications. I show that when one analyzes his corpus as a whole, a very different image emerges from the one that appears if one focuses its study only on this area of his work. This analysis, decentered from the axis around which the work of Hacking has traditionally been understood, is the basis of my proposal for this book. To defend it, I articulate the items dealt with in previous chapters and I close the circle I open in the first chapter, going back to the proposal of a reticular structure in Hacking's work and the place Michel Foucault occupies in this web, which I define as the texture, the always underlying thread that runs across in the loom where the cloth is woven. I claim that it is the notion of conditions of possibility for the emergence of concepts and scientific objects, inherited and at the same time different from Foucault's, what impregnates Hacking's whole corpus.

Keywords Texture in Ian Hacking's work · Analysis of conditions of possibility · Michel Foucault's influence on Hacking's work · Underlying interests in Hacking's thinking · Scientific realism

---

Hacking alludes to this expression in an interview, apropos how much he had learnt by reading Foucault's books and comparing his writing style with that of the French philosopher (Madsen et al. 2013: 43)

M. L. Martínez Rodríguez, *Texture in the Work of Ian Hacking*, Synthese Library 435, https://doi.org/10.1007/978-3-030-64785-8_7

7.1 Scientific Realism in Hacking's Work Viewed as a Whole

Within the analytic tradition, the work that has given Ian Hacking more visibility and consideration, or at least the book that has brought more readers to him, is *Representing and Intervening.* This was, in particular, my own experience. I approached Hacking's thought through this book and its proposal about the styles of scientific reasoning. It was after a while working with this topic that I discovered the rest of this work, which unfolded for me a much wider landscape, mainly in relation to the philosophy of the human sciences. This allowed me to ascertain that, whereas Hacking's best-known book is this one of 1983, at that time, and even before, Hacking worked and reflected about the field of the human sciences, and even before he already had in mind, at least as an outline, his notion of styles of scientific reasoning.

It turns out then that when we analyze Hacking's work as a whole what emerges is a very different image from the one that one may have by focusing only or mainly on *Representing and Intervening* and some related articles. Hacking has devoted much less work and in shorter periods of time to the analysis of the natural sciences if we compare it to the other topics he deals with. Whereas we can say that he published this book and some other articles on the natural sciences between 1980 and the mid-90s, after 1975, when he published *The Emergence of Probability*, he continued to work on issues related to the human sciences and styles of scientific reasoning until today.

This observation has led me to wonder, among other questions: which have been – or continue to be— the interests that underlie Hacking's *oeuvre* as a whole? What are the grounds of his interest in these topics and how has he navigated through them?

Hacking himself in an interview mentioned at the beginning of this book (Álvarez Rodríguez 2002:1) names some of the parts of the network that makes up his work. He mentions probability, making up people and Michel Foucault. At first, I thought that the absence of scientific realism was due to the fact that he was only giving a few examples. However, in the light of my research, I think that there might be a different answer. Hacking does not mention realism and his work on the natural sciences because that is not the core of his work but just an offshoot of a much more general and wide-ranging research on the historical and situated conditions of possibility for the emergence of objects and concepts in general.

In *Scientific Reason,* Hacking reminds us of the structure of his book of 1983, divided into two parts: representing and intervening. The second includes an argument for a realism about theoretical unobservable entities. Regarding this, Hacking wonders today: is realism important? No. Was it important then (in 1983) and then he changed his mind? No. He goes on to evoke the following lines from the beginning of his book:

> Disputes about reason and reality have long polarized philosophers of science [...] Is either kind of questions important? I doubt it. We do want to know what is really real and what is

truly rational. Yet you will find that I dismiss most questions about rationality and am a realist on only the most pragmatic of grounds (Hacking 1983: 2, 2009: 146).

Hacking says he still has doubts about the importance of these questions 26 years after the publication of his book. His realism had a pragmatic basis. The book was about experiments and its aim was to vindicate its independence with regards to theory or, in other words, to defend that the experiment has a life of its own. At the time, philosophers in general were not interested in experiments and scientific realism was in vogue. Hacking wanted to open a window to be able to speak about the experiment. To that end, he used the controversy about scientific realism as a strategy for his work. In this sense, there is no incompatibility between his experimental argumentation in favor of scientific realism and his declared lack of interest in the debates around realism. But this also strengthens my idea that the topic that has gained Hacking more recognition at least within Anglo-Saxon philosophy of science is not the one that has interested him the most.

7.2 The Analysis of Conditions of Possibility as the Main Interest of Hacking's Work

In the reticular structure of Hacking's work, I have identified four nodes: (1) styles of scientific reasoning/thinking & doing, (2) probability, (3) making up people, and (4) experimentation and scientific realism. Each one of them gives place to interwoven research lines, so that the probabilistic style is an exemplar of style of scientific thinking & doing and this is in its turn an attempt to solve and generalize questions that emerge from case studies such as probability and statistics. But styles of scientific thinking & doing are also a source of making up people through classifications that emerge in the census or population statistics, for example. They are, besides, a historic and situated condition of possibility for the emergence not only of new kinds of people and objects but also realist debates which, far from being debates about realism and anti-realism in general, fall within the limits of each style, since each one of them proposes the existence of certain objects. Styles of scientific thinking & doing deal with problems of categorization and nomenclature that Hacking develops in his works on kinds –natural and human—and in general terms of doing and intervention. This central place of the style of scientific thinking & doing is what has led me to conceive it as a basal node or core from which all the others derive and become related, and the reason why I have presented it in the first place.

My aim here is to underscore that, against what is often thought, the main axis around which to understand Hacking's work is not his concern about realism and the natural sciences. Hacking's work as a whole can be better structured and understood around the axis of his interest in the analysis of the historical and situated conditions of possibility for the emergence of concepts and objects which underlies his *oeuvre*. Perhaps the clearest and earliest example of this interest can be found in

The Emergence of Probability, where he shows how the concept of probability emerges around 1660 as a result of a mutation/crystallization of earlier renaissance conceptions. In that same year, 1975, *Why Does Language Matter to Philosophy?* deals with the analysis of language. In 1983, *Representing and Intervening* analyzes, from the starting point of his emphasis of experimentation and manipulation, the emergence of laboratory phenomena. These phenomena, such as the laser, do not exist until they are made in the laboratory. Seven year later, in *The Taming of Chance*, Hacking shows which conditions allow for the emergence of what he calls *making up people*. The growth of statistical analysis, the avalanche of census numbers, clearly shows how from the starting points of new categorizations it is possible to construct new kinds of people. It is possible to construct objects that have a historical ontology. "Historical Ontology", Hacking's article on this topic which names the collection of articles published in 2002, has as its central axes Foucault and the uses that philosophy makes of history, but also the books *Rewriting the Soul* and *Mad Travelers*, where he studies the concept of multiple personality disorder and its relation with memory and child abuse and the conditions of possibility for the emergence of the so-called compulsive travelers or *fugueurs*, respectively. Finally, Hacking presents the styles of scientific thinking & doing as providing the conditions that allow for the emergence of concepts, objects, and classes, characteristic of each style. It is in this sense, besides, that I have identified the notion of style of scientific thinking & doing as a central node or project, insofar as it provides these conditions.

Hacking has remarked on numerous occasions the influence of Michel Foucault's works on his thought, and he has said that Foucault exemplifies what philosophy is for him: "[…] *a way of analyzing and coming to understand the conditions of possibility for ideas*" (Hacking 1981: 76). It is important to remember here what was said in the third chapter of this book: Hacking is not interested in the transcendental conditions of possibility in Kant's way, and he even goes beyond Foucault in his interest in the pre-history of concepts. Hacking is interested in the conditions of possibility—theoretical, historical, pragmatic, semantic, social. In his works he analyzes how concepts and objects not only become possible but also how they are historically altered.

7.3 Michael Foucault: The Texture in Hacking's Work

It is this notion of the conditions -in Hacking's case, historical and situated- of possibility, inherited but at the same time different from Foucault's, that impregnates Hacking's whole work. This is what motivates my proposal of pointing out Foucault's influence as the thread that runs across Hacking's work.

In answering the interviewer's question, Hacking places Michel Foucault as one more section in the network of his work. I believe that the French philosopher occupies a very different place. I defend that Michel Foucault's influence on Hacking's work, rather than representing a portion of the network, constitutes the texture that weaves together the whole structure, which emerges with greater or lesser visibility, with greater or lesser strength, but is always present.

Foucault is clearly present in Hacking's analyses of the human sciences, of styles of scientific thinking & doing, of probability. But he is also present in Hacking's analyses of the natural sciences, at least in the sense that Foucault stimulated his interest in investigating the conditions of possibility for the emergence of the concepts and objects that belong to this realm of the sciences. It is worth remarking here that Hacking never abandons his analytical filiation, although it might be more apparent in some nodes than in others. His philosophy of the natural sciences is clearly an example of the influence of the tradition in which Hacking was trained and which he always acknowledges.

I also chose the term "texture" because I understand that beyond Foucault's punctual influence through some of his books and particular notions in so many books and notions of Hacking's, there is a more general influence that underlies his whole work as a thread that runs through that network or texture that is Hacking's work. It could be said, in some sense, that Foucault, and in particular his concern for the search of conditions of possibility, is the one who inspires or stimulates Hacking's interest and research through the analysis of the conditions of possibility of the emergence of objects and concepts.

But Hacking has also been influenced by Foucault through his methodology. As I have remarked, the archaeological methodology can be perceived, albeit in different degrees, since Hacking's earliest works, and is complemented in others, such as *The Taming of Chance* or *Rewriting the Soul*, with his genealogical contribution.

Hacking is a disseminator of Foucault's work. He is one of the main thinkers responsible for Foucault being known in the field of Anglo-Saxon/analytic philosophy of science. In the spring of 1974 Hacking gave a series of conferences about some of Foucault's works, and as he himself comments in "*Les mots et les choses*, forty years on" (2005), it is said that on this occasion a colleague of his told a visitor: "*If you wonder why the bookshops have copies of Foucault in their front windows, it is all Hacking's fault*" (Hacking 2005: 3).

However, Hacking is not just a disseminator of the ideas of the French philosopher. On the contrary, he admits how much he has learned by reading his books and using some of his ideas. He adopts and adapts Foucault's notions for his own purposes. As we have seen, in some cases he goes beyond Foucauldian thought, and in others he stays behind. In any case, Hacking does not think of himself as Foucault's disciple.

As an English-speaking philosopher influenced by the French philosopher, Hacking respects his origins in the analytic tradition but does not accept its restrictions. As he himself says (1990: 70–71), he was trained as an analytic philosopher with an emphasis on philosophical logic, someone whose mind was shaped by Frege, Moore and Russell, and in this sense, he is interested in the clarity of concepts. But Hacking goes beyond his original analytic tradition and considers that, regardless of their clarity, it is not possible to use concepts correctly if their trajectory is not known. Hacking is not an analytic philosopher who, in spite of Wittgenstein, deals with concepts as if they exist previously to any use, as if their identification did not depend on their interconnections, of what can be done with them in reality. On the contrary, concepts are situated words. In this respect, not only does Hacking theorize about concepts being words in their sites, but his philosophical analyses do not take concepts in an abstract and non-temporal way. On the contrary, they are approaches from a historical perspective that might help understand the actuality of the concept and eliminate possible confusions.

As an illustration, and to add one more example to what we have already shown *in extenso* throughout this book regarding the influence of Foucault's ideas on Hacking's thought, I propose to take a moment to analyze a text that was not discussed in depth previously. In 1975, in *Why Does Language Matter to Philosophy?*, when presenting the way in which language has mattered to philosophy, a clear historical dimension can be perceived in Hacking's analysis, which is generally missing in analytic philosophers. Even though the aim of that text is to analyze what happens in the present, it requires a historical perspective to show when language has mattered to philosophy. This interest in context is due, according to the philosopher, to the impact of the reading of *The Order of Things*. Hacking deems necessary to adopt a historical perspective that can contribute to an understanding of the present, to understand how and why we have reached our present conceptions, on the basis of an account of their origin. To this end, he attempts to apply to the sequence of changes in the practices of western philosophy regarding the relation between cognitive representations and the world, a periodization similar to the Foucauldian one. Just like Foucault analyzes in his book (1) the Renaissance, (2) what he calls the Classical Period (17th and 18th cent.), and (3) Modernity (nineteenth and twentieth century), Hacking will divide his analysis in: (1) heyday of ideas (17th cent.); (2) heyday of meaning (beginning of the 20th cent.) and (3) heyday of sentences (after the mid-20th cent.).

But the division of the analysis in three periods is not the only common element in these two books; they also share an interest in language, which is one of the main topics of Foucault's philosophical reflection.

In his book, Hacking analyzes in the first place the heyday of ideas, during the seventeenth century, when the mental language has priority over the public language. The ideas of reality are the result of the action of the experience on the ego, and simultaneously, the cause of this experience. There is no external knowledge except through the ideas that are within ourselves. It is the Cartesian ego that fixes the framework. The ego capable of contemplating what is inside man and at the same time of taking into consideration what lies outside. According to Hacking, during this first heyday, there is no theory of meaning in the sense that it is conceived at present. There is no concern for meaning in Fregean terms. The representatives of this heyday worked on something structurally similar to present-day problems, but the place that now corresponds to the public was then occupied by the private.

In this particular period there is a clear influence of Foucault's treatment, mainly in Chap. 4 of *The Order of Things*, mainly about general grammar and the proposal of the philosophers of Port-Royal, who were vividly concerned about grammar, to the point of considering their contributions in this field more important than those related to logic. The problem of general grammar – both for Hacking and for Foucault—has to do with articulation and it is in this sense that the former remarks: "*The problem of general grammar is to explain how articulated language effects the representation of a non-articulated part of the world*" (Hacking 1975: 87).

In other words, the central problem of the metaphysics of that time, even though it derives from the doctrine of ideas, is not inherent to the ideas themselves. It has to do with the relation between ideas, words and things; with the question of the actual workings of representation by means of words when they present themselves in an articulated sequence but the things in the world do not.

For Foucault, grammar is general precisely in the sense that "*[…] it makes language visible as a representation that is the articulation of another representation, it is indisputably 'general'; what it treats of is the interior duplication existing within representation*" (2005: 101, 1966: 106).

However, in my view, *The Order of Things* is not the only influence that can be detected in this period. Also *The Birth of the Clinic* left an imprint on it. In pointing out how the Cartesian world is completely visual, Hacking (1975: 32–33) practically paraphrases Foucault when he remarks that for Descartes and Malebranche seeing with the eyes is perceiving with the mind. To perceive is to turn the object transparent. At the end of the eighteenth century, continues Hacking paraphrasing Foucault, this kind of perception was replaced by our kind of vision. Objects become opaque and resist physical light instead of giving way to mental light. Hence seeing is the passive clash between luminous rays in opaque physical objects, impermeable, which are themselves passive and indifferent regarding the subject of the observation.

In any case, Hacking will later regret (Vagelli 2014: 261) not including in this first chapter Foucault's thoughts about language, about the importance of the way in which the notion of a private language became obsolete. He later published his reflections on this question in his work "How, Why, When and Where did Language Go Public?", included in *Historical Ontology.*

During the heyday of meaning, the second stage identified by Hacking -and in his view massively influenced by Foucauldian thought (Vagelli 2014: 261)- what is claimed is the need that behind comprehensible sentences there exist meanings that are the true carriers of belief and knowledge. There is something below the level of what is said: what is meant. Meanings make public discourse possible. Following Frege, Hacking claims that a theory of meaning is a theory about the possibility of public discourse.

The analysis of this period is the occasion for Hacking to point out critically two ideas he will return to in later works. In the first place, with the heyday of meanings, it is thought that it is possible to decide substantive philosophical problems by contemplating meanings. This leads to a new kind of idealism that, to avoid the solecism inherent in the word "idea", could be called "lingualism". Secondly, the idea that philosophy is the slave of grammar. Mistaken notions about language leads to bad philosophy. A better and more analytical language is needed to decode the truth. But real philosophy cannot be the slave of grammar; on the contrary, far from being autonomous and constituting the substance of ontology, grammar responds to the world and to what is in it (Hacking 1975: 172–173).

The last period, the heyday of sentences, beings with the failure of the verificationist process and the frequent doubts about the precision of meaning. Despite (and respecting) their differences, Hacking associates Quine and Feyerabend during this period, understanding that both object to the elements of a positivist methodology, but still promote a similarly positivist movement, moving away from meanings and converging on sentences. Sentences have replaced ideas. Knowledge has become sentence-like. This topic will be taken up and criticized by Hacking in his book of 1983, where he claims that in the philosophy of that time there was a tendency to substitute observations by linguistic entities and that if we want a comprehensive description of scientific life we should, contrary to proposals such as Quine's, not speak of observational sentences but of observation (1983: 181).

At the end of *Why Does Language Matter to Philosophy?* Hacking refers once again to Foucault by pointing out that recognizing an autonomous, essentially sentence-like knowledge, introduces new objects of research, new domains of inquiry. The French philosopher has postulated sentence-like discourses that exist in different places and times. These discourses are not identified because of what they mean but because of what is actually said in certain places and under the aegis of certain institutions.

Along these lines, some years later, in his article "Five Parables", Hacking analyzes, as we have seen, the relation between philosophy and its past; he reflects about the uses of history in philosophy of science and claims that history plays a much more important role than merely removing linguistic confusions. This other role, central to differentiating between the objects of the natural and the human sciences and to the contrast between the creation of phenomena and making up people, has to do with the fact that whereas the objects of the human sciences are constituted by a historical process, those of the natural sciences are not. Despite his analytical training, he sees no inconsistency between his analytic instincts and his loyalty to the Lockean imperative: to seek the understanding of thoughts and beliefs on the basis of an account of how they originate. For Hacking concepts have a history, the objects of the human sciences have a history, the ways of telling the truth have a history, the ways of researching have a history. Conditions of possibility have a history. Foucault, for whom also the subject of knowledge, the relation between subject and object, and truth, have a history, analyzes how we come to concepts and how they vary. Hacking reflects about this and also about why a new concept might be preferred to a previous one.

It has been a declared aim of Hacking's to extend a bridge between analytic and continental philosophy without losing the power of either, showing – as he attempts to do in *The Taming of Chance*—that analytic philosophy and historical sensibility do not need to be antithetical, that they can be convergent.

As Hacking already remarked, a conciliation between both is neither desirable nor necessary because they are not antithetic traditions. "*I have no desire to make peace among different traditions. Attempts to reconcile continental and analytical philosophy are at best bland, lacking the savoir or pungency of either*" (Hacking 2002a: 51–52).

This is why Hacking attempts to read old philosophical texts in a new way. A way that leads to the analysis of words in their places in order to understand how we think and why we seem compelled to think in a certain way. Hacking believes that this Foucauldian history of the present is important not only to understand concepts in general but also to understand the root of their problematic nature. It is necessary to take into account the prehistory of problematic concepts and what made them possible to grasp the nature of philosophical problems. In any case, even though this understanding might explain problems, it does not manage to prevent them from disturbing us; in other words, it will not make them disappear. Hacking has insisted that in Foucault's work there is a new and unique attempt to make evident and explicit the real and deep sources of the historiographic wonder in front of a reading of texts from previous times that have the pretension to systematic knowledge but that cannot be honestly apprehended with local conceptual resources. To analyze words in their places allows for a more objective articulation of the knowledge of periods other than the present, because it reveals the deeper-order changes that make them possible, and the real historical articulation with obsolete conceptual frameworks and practices.

Gianni Vattimo has remarked in the preface to *Analíticos y continentales, Guía de la filosofía de los últimos 30 años* that, in many senses, the separation or opposition between the analytic and continental styles of thinking is perhaps the question that summarizes the characteristic problem of present-day philosophy (2000: 14). In this sense, Hacking's proposal constitutes a particular case of linking these two great styles, insofar as in his work we find clearly conjugated the features of his analytic philosophical origin and others, derived from continental philosophy, of a humanistic orientation, that considers history determining and thinks logic as art of the logos or discipline of the concept rather than as calculation or computation (Ibid: 24) and whose texts exhibit a style of argumentation and thought closer to the humanistic disciplines, where history appears as an important methodological factor (Ibid: 36).

References

Álvarez Rodríguez, A. (2002). Entrevista con Ian Hacking. *Cuaderno de Materiales*, 17. http://www.filosofia.net/materiales/num/num17/Hacking.htm

Foucault, M. (1966). *Les mots et les choses. Une archéologie des sciences humaines*. Paris: Gallimard.

Foucault, M. (2005). *The order of things. An archaeology of the human sciences*. London and New York: Routledge.

Hacking, I. (1975). *Why does language matter to philosophy?* Cambridge: Cambridge University.

Hacking, I. (1981). The archaeology of Michel Foucault. In I. Hacking (2002a), *Historical ontology* (pp. 73–86). London: Harvard University.

Hacking, I. (1983). *Representing and intervening*. Cambridge: Cambridge University.

Hacking, I. (1990). Two kinds of new historicism for philosophers. In I. Hacking (2002), *Historical ontology* (pp. 51–72). London: Harvard University.
Hacking, I. (2002a). *Historical ontology*. London: Harvard University.
Hacking, I. (2002b). *L'émergence de la probabilité*. Paris: Seuil.
Hacking, I. (2005). *Les Mots et les Choses*, forty years on. *For humanities center* (pp. 1–24). Columbia University.
Hacking, I. (2009). *Scientific reason*. Taipei: National Taiwan University.
Madsen, O., Servan, J., & Øyen, S. (2013). "I am a philosopher of the particular case": An interview with the 2009 Holberg prizewinner Ian Hacking. *History of the Human Sciences, 26*(3), 32–51.
Vagelli, M. (2014). Ian Hacking. The philosopher of the present. *Iride, 27*(72), 239–269.
Vattimo, G. (2000). Prefacio. En F. D'Agostini (2000), *Analíticos y continentales* (pp. 13–17). Madrid: Cátedra.

Epilogue

Abstract Finally, in the *Epilogue*, I claim that Ian Hacking's proposal for the human sciences, mainly his notions on how to make up people, looping effect and interactive classifications, not only can be complementary of Michel Foucault's thought, but that this complementarity can give as a result a vision that goes beyond both projects considered separately, and offer a better explanation of the objects of the human sciences and their behavior. This is the case insofar as Hacking's most concrete proposal would manage to complete Foucault's more abstract analyses, showing one of the mechanisms by means of which a relation is established between practices, institutions, and discourse on the one hand, and people and their behaviors in daily life on the other.

Keywords Making up people · Looping effect · Interactive kinds · Complementary approaches of Ian Hacking and Michel Foucault · Human sciences

I would like to conclude with a specific analysis of a particular point where Hacking's proposal can not only be complementary of Foucault's thought, but also result in a vision that goes beyond both projects considered separately and offer a better explanation of the objects of the human sciences and their behaviors: his proposal of making up people, interactive kinds and looping effect.

There are subjects on which Hacking remarks he did not go as far as Foucault. In 2010 he claims: "*I have learned much from Michel Foucault, but I do not want to unmask anything*" (April 21: 4), confirming somehow his previous remarks that his historical ontology is a contraction of Foucault's insofar as "*It lacks the political ambition and the engagement in struggle that he intended for his later genealogies*" (Hacking 2002: 5). In any case, according to Hacking, his is a tepid activism

M. L. Martínez Rodríguez, *Texture in the Work of Ian Hacking*, Synthese Library 435, https://doi.org/10.1007/978-3-030-64785-8

compared with Foucault's who was an activist on very practical aspects such as prison reform.

On the other hand, it must be taken into account that, beyond the fact that Hacking might genuinely not have had interest in committing himself politically as Foucault did, he had to practice the Foucauldian methodology in ways that were acceptable in the realm of analytic philosophy. An illustrative example in this sense is the subject of power, implicitly suggested in numerous occasions in Hacking's work but never developed systematically, as did Foucault. Hacking does not analyze the relation power-classification or how power is embedded in the process of interaction between the classification and the classified individuals. Neither do his works show an explicit thematization of the intentionality of classifications and their consequences, as can be seen in Foucault's work. Nevertheless, as I have already remarked, the fact that he decides not to work on this subject does not mean that this has not influenced his work and is not visible in it. This can be perceived not only in his texts on the human sciences but also in his consideration that Foucault's works are useful to grasp the interrelations between power and knowledge (classifications) that constitute human beings. Hacking developed in *The Taming of Chance* the subject of control and quantification when he analyzed how statistics were not a mere report but rather "*a direct and visible element in the exercise of power*" (Hacking 1990: xi), and that his collection generated a great bureaucratic machine that "*may think of itself as providing only information, but it is itself part of the technology of power in a modern state*" (Hacking 1981b: 181). Statistical technologies gave place to the study of populations and transformed them in objects of knowledge. The collection of massive quantities of data was a necessary condition to reconceptualize patterns and to define norms within a multiplicity and, ultimately, to control and modify social practices.

Hacking also devoted a considerable part of his article "The Archaeology of Michel Foucault" to the analysis of the French philosopher's treatment of power. It is on the basis of these ideas that, even though Hacking does not theorize about power, I devote a chapter to show how it plays a fundamental role or is embedded in Hacking's proposal about classifications, the looping effect and making up people. In analyzing not only how the elements Foucault considers essential in a power relation – 1. The person classified is recognized by the expert and considered until the end as a person that acts, that reacts, and 2. Confronted with this classification, for the person classified a whole new field opens up of responses, reactions, results and possible inventions-, become visible in the relation between the human kinds and their members and behaviors, but also that in this relation there can be observed the main features that Foucault sees in power. Power is a way in which certain actions modify others; it circulates among experts and the people classified, it is exercised not only from the top but also from the bottom; power produces, creates new realities; power generates resistance, a resistance that is as inventive, mobile and productive as power itself; etc.

Despite Hacking's claims that he does not set out to unmask anything, I believe that insofar as his work has as an aim to be critical, in the sense of showing the contingency of certain classifications, its effect can be the possibility that individuals may become more aware of the dynamic relation they have with the labels with which they are described not only by others but by themselves. Therefore, even though Hacking is not intentionally interested in a political sense to modify anything, I believe that his work opens an interesting field of scientific and ethical discussion on this point. Classification as value-laden classes entail, at least, a moral problem. Suffice it to mention, as an example, the interest in elucidating the epistemic-ethical relations established on the basis of the idea that a way of knowing implies a way of classifying, which in its turn entails a series of inherent values. Commenting on the case of Pierre Rivière analyzed by Foucault, Hacking points out that, perhaps for the first time, a multitude of experts were in court theorizing about the supposedly mad assassin. The categories assigned to him determined what had to be done with him. This is a small way in which knowledge is power. It was not so much the facts concerning Pierre as the possibility of thinking about him in a certain way that determined his destiny (1981a: 80).

The notions of making up people and the looping effect, even though they were inspired and stimulated by Foucault's works, deal in my opinion with aspects that he did not develop in depth and, in this sense, I consider that Hacking goes beyond Foucault and his proposal can complement that of the French philosopher, resulting in a more complete and ambitious proposal.[1]

A frequent criticism of Foucault is that his analysis tends to leave out human agency. There is something missing in his approach, an understanding of how the forms of discourse become part of the lives of common people or how they are institutionalized and become part of the structures of the institutions where they work. The proposed classifications exist only within a matrix of practices, and therefore a new way of describing people creates not only new ways of being but new ways of choosing. The space of possible and real action is not determined only by physical, social and opportunity limits but also by the ways in which people are conceptualized and, on this basis, manage to realize who they are and who they can be in each specific historical moment. As Hacking points out about *History of Madness*, insane people are classified, dealt with and separated by means of systems of our own creation. Our institutions create the phenomena in terms of which we regard insanity. Foucault's book suggests a quasi-Kantian narrative where our experience of the insane is a phenomenon conditioned by our thought and our history, even though there is a thing in itself that can be called madness and is incorruptible (1981a: 75).

Foucault establishes conditions for institutional forms and analyzes their mutations, but he does not study how these historic landscapes and their influence and determination become part of people's daily lives. He does not do a history of behaviors or of ideas. His aim is to do a history of the conditions under which

[1] This analysis could be completed by the studies of Erving Goffman on daily social interaction. Goffman analyzed local and idiosyncratic incidents that lead us from bottom to top, showing how changes are not deliberately induced by the control system but rather that they take place in the presence of other people and due to that presence. It is a question of the gestures, postures, words, that each person inserts his or herself, intentionally or otherwise, in the situation.

certain behaviors and statements appear. Likewise, he does not do a history of private life, but rather a history of subjectivation as the condition of private life as a whole.

On this point, I consider that Hacking's proposal can be visualized as an improvement in this aspect. In some sense, Hacking's analyses complement or concretize what the French philosopher proposes in more abstract and general terms. Foucault explains how madness, the mentally insane and the institutions around them have changed. Hacking, for his part, explains in more concrete terms how the interaction between these more general elements and the mentally insane as an individual, as a member of a class, takes place. He explains one of the mechanisms (the relation classification-classified) by means of which these changes take place, a mechanism that is at the same time cause and consequence of the changes: the looping effect between the classifications and the classified members, that modifies the behavior of the mentally insane individual and that which actually changes madness and everything it implies and surrounds it. The classified persons, as we have seen, are not passive victims of a classification but active agents that interact with the categorization and create something new. They react to the classification and produce a change in the classifications themselves, the institutions and the forms of discourse that must now be adapted to describe the new members of the class. Foucault always rebelled against the images of top-down power. It is the subjects that choose the specific forms in which they will be subjects. Hacking's idea that the persons classified react to a classification and produce the looping effect, exercising power on the expert and everything that surrounds the categorization, takes up magnificently this concern of Foucault's, because it shows a way in which power is exercised bottom-up.

Hacking's looping effect can explain how it is possible that a person may adopt certain roles at a specific point in time and place. Foucault shows the ways in which it is possible to understand what is possible, what can be said, what is said, what is meaningful. Hacking shows us how people, in their daily lives, by means of their interaction with the classification and the whole matrix it implies, incorporate these possibilities and impossibilities as part of themselves and how, by reacting to them, finally lead to a modification of what is possible, what can be said, what is said and what is significant.

Foucault has said: "*Deep down, the question I have asked myself was not so much knowing what was going through the minds of the diseased, but what went on between them and the physicians*" (Foucault 2012: 30; 1994. Vol 3: 369).

I consider that Hacking's proposal gives a satisfactory reply to the French philosopher's question: the classification and the consequent looping effect is one of the things that goes on between them.

References

Hacking, I. (1990). *The taming of chance*. Cambridge: Cambridge University.
Hacking, I. (2002). *Historical ontology*. London: Harvard University.
Hacking, I. (1981a). The archaeology of Michel Foucault. In I. Hacking (2002), *Historical ontology* (pp. 73–86). London: Harvard University.
Hacking, I. (1981b). How should we do the history of statistics? In G. Burchell, C. Gordon, & P. Miller (Eds.) (1991), *The Foucault effect. Studies in governmentality* (pp. 181–195). Chicago: Chicago University.
Hacking, I. (2010, April 21). Lecture I. Methods, objects, and truth. [Unpublished]. México: UNAM.
Foucault, M. (1994). *Dits et écrits* (Vol. 1–4). Paris: Gallimard, édités par D. Deferí & F. Ewald.
Foucault, M. (2012). *El poder, una bestia magnífica: Sobre el poder, la prisión y la vida*. Buenos Aires: S. XXI.

Bibliography

Agassi, J. (2005). Back to the drawing board. *Philosophy of the Social Sciences, 35*(4), 509–518.

Ali Khalidi, M. (1998). Natural kinds and crosscutting categories. *The Journal of Philosophy, 95*(1), 33–50.

Balbier, E., Deleuze, G., Dreyfus, H. L., Frank, M., & Glúksmann, A. (1999). *Michel Foucault, filósofo*. Barcelona: Gedisa.

Bueno, O. (2018). Estilos de Raciocínio nas Ciências e nas Artes. En S. Chibeni, *Filosofía e Historia de la Ciencia en el Cono Sur. Selección de trabajos del X Encuentro de la Asociación de Filosofía e Historia de la Ciencia del Cono Sur* (pp. 27–39). Córdoba: AFHIC.

Castro, E. (2004). *El vocabulario de Michel Foucault. Un recorrido alfabético por sus temas, conceptos y autores*. Buenos Aires: Universidad Nacional de Quilmes.

Castro, E. (2011). *Lecturas foucaulteanas. Una historia conceptual de la biopolítica*. La Plata: Unipe.

Castro, J. (2017). *Las relaciones entre estilos de razonamiento y prácticas científicas como eje central de un proyecto de epistemología histórica*. México: Ciencia Nueva. Doctorados UNAM. http://www.ciencianueva.unam.mx/handle/123456789/164

Chandler, J., Davidson, A., & Harootunian, H. (1994). *Questions of evidence. Proof, practice, and persuasion across the disciplines*. Chicago: University of Chicago.

Cisney, V., & Morar, N. (2015). *Biopower. Foucault and beyond*. Chicago: University of Chicago.

Couzens Hoy, D. (1999). *Foucault. A critical reader*. Oxford: Blackwell.

Crombie, A. C. (1995). Commitments and styles of European scientific thinking. *History of Science, 33*(100), 225–238.

D'Agostini, F. (2000). *Analíticos y continentales*. Madrid: Cátedra.

Daston, L. (1994). Historical epistemology. In J. Chandler, A. Davidson, H. Harootunian (1994), *Questions of evidence. Proof, practice, and persuasion across the disciplines* (pp. 282–289). Chicago: University of Chicago.

Daston, L. (2000). *Biographies of scientific objects*. Chicago: University of Chicago.

Deleuze, G. (2005). *La isla desierta y otros textos. Textos y entrevistas (1953–1974)*. Valencia: Pre-Textos.

Deleuze, G. (2013). *El saber: curso sobre Foucault I*. Buenos Aires: Cactus.

Deleuze, G. (2014a). *El poder: curso sobre Foucault II*. Buenos Aires: Cactus.

Deleuze, G. (2014b). *Michel Foucault y el poder. Viajes iniciáticos I*. Madrid: Errata Naturae.

Deleuze, G. (2015a). *La subjetivación: curso sobre Foucault III*. Buenos Aires: Cactus.

Deleuze, G. (2015b). *Foucault*. Buenos Aires: Paidós.

M. L. Martínez Rodríguez, *Texture in the Work of Ian Hacking*, Synthese Library 435, https://doi.org/10.1007/978-3-030-64785-8

Desrosières, A. (2006). Les recherches de Ian Hacking sur l'histoire des usages des probabilités et des statistiques dans le raisonnement inductif. *Journ@l Electronique d'Histoire des Probabilités et de la Statistique, 2*(1), 1–8.

Drabek, M. L. (2010). Interactive classification and practice in the social sciences: Expunging Ian Hacking's treatment of interactive kinds. *Poroi, 6*(2), 62–80.

Elden, S. (2017). *Foucault. The birth of power*. Cambridge: Polity.

Elwick, J. (2012). Layered history: Styles of reasoning as stratified conditions of possibility. *Studies in History and Philosophy of Science, 43*, 619–627.

Foucault, M. (1963). *Raymond Roussel*. Paris: Gallimard.

Foucault, M. (1971). *L'ordre du discours*. Paris: Gallimard.

Foucault, M. (1976–1984). *Histoire de la sexualité* (Vol. 1–3). Paris: Gallimard.

Foucault, M. (1978). *A verdade e as formas jurídicas*. Rio de Janeiro: Pontificia Universidade Católica do Rio de Janeiro.

Foucault, M. (2000). *Power. Essential works of Foucault 1954–1984* (Vol. 3). New York: The New Press.

Foucault, M. (2003). *Le pouvoir psychiatrique. Cours au Collège de France (1973–1974)*. Paris: Seuil-Gallimard.

Foucault, M. (2004). *Naissance de la biopolitique. Cours au Collège de France (1978–1979)*. Paris: Seuil-Gallimard.

Foucault, M. (2006). *Madness and civilization: A history of insanity in the age of reason*. New York: Vintage Books.

Foucault, M. (2008a). *Introduction to Kant's anthropology*. Los Ángeles: Semiotexte.

Foucault, M. (2008b). *Introduction à l'Anthropologie de Kant*. Paris: Librairie Philosophique J. Vrin.

Foucault, M. (2008c). *Un diálogo sobre el poder y otras conversaciones*. Buenos Aires: Alianza.

Foucault, M. (2009). *Le courage de la vérité. Le gouvernement de soi et des autres II. Cours au Collège de France (1983–1984)*. Paris: Seuil-Gallimard.

Foucault, M. (2013a). *¿Qué es usted, profesor Foucault?* Buenos Aires: S. xxi.

Foucault, M. (2013b). *La inquietud por la verdad*. Buenos Aires: S. XXI.

Foucault, M. (2014). *Las redes del poder*. Buenos Aires: Prometeo.

Foucault, M. (2015). *Saber, historia y discurso*. Buenos Aires: Prometeo.

Foucault, M., y Chomsky, N. (2006). *La naturaleza humana: justicia versus poder. Un debate*. Buenos Aires: Katz.

Gabilondo, Á. (1990). *El discurso en acción. Foucault y una ontología del presente*. Madrid: Anthropos.

Gutting, G. (2001). *French philosophy in the twentieth century*. Cambridge: Cambridge University.

Gutting, G. (Ed.). (2005). *The Cambridge companion to Foucault*. Cambridge: Cambridge University.

Hacking, I. (1967). Slightly more realistic personal probability. *Philosophy of Science, 34*(4), 311–325.

Hacking, I. (1980). Is the end in sight for epistemology? *Journal of Philosophy, 77*, 579–588.

Hacking, I. (1981). *Scientific revolutions*. Oxford: Oxford University.

Hacking, I. (1983). Nineteenth century cracks in the concept of determinism. *Journal of the History of Ideas, 44*(3), 455–475.

Hacking, I. (1984a). Wittgenstein rules. *Social Studies of Science, 14*, 469–476.

Hacking, I. (1984b). Self-improvement. In I. Hacking (2002), *Historical ontology* (pp. 115–120). London: Harvard University.

Hacking, I. (1987). The inverse Gambler's fallacy: The argument from design. The anthropic principle applied to Wheeler universes. *Mind, 96*(383), 331–340.

Hacking, I. (1988a). On the stability of the laboratory sciences. *The Journal of Philosophy, 85*(10), 507–514.

Hacking, I. (1988b). The sociology of knowledge about child abuse. *Noûs, 22*, 53–63.

Hacking, I. (1988c). Telepathy: Origins of randomization in experimental design. *Isis, 79*, 427–451.

Hacking, I. (1989a). Philosophers of experiment. *PSA, 1988*(2), 147–156.
Hacking, I. (1989b). The life of instruments. *Studies in History and Philosophy of Science, 20*(2), 265–270.
Hacking, I. (1989c). Extragalactic reality: The case of gravitational lensing. *Philosophy of Science, 56*, 557–581.
Hacking, I. (1990). [The unity and diversity of probability]: Comment: In praise of the diversity of probabilities. *Statistical Science, 5*(4), 450–454.
Hacking, I. (1991a). On boyd. *Philosophical Studies, 61*(1–2), 149–154.
Hacking, I. (1991b). Artificial phenomena. *The British Journal for the History of Science, 24*(2), 235–241.
Hacking, I. (1992). Multiple personality disorder and its hosts. *History of the Human Sciences, 5*(2), 3–31.
Hacking, I. (1993). On Kripke's and Goodman's uses of Grue. *Philosophy, 68*(265), 269–295.
Hacking, I. (1994). Aristotle meets incest-and innocence. In J. Chandler, A. I. Davidson, & H. Harootunian (Eds.), *Questions of evidence. proof, practice, and persuasion across the disciplines* (pp. 470–477). Chicago: Chicago University.
Hacking, I. (1995a). Comments on Zeidler & Sobczynska's paper. *Foundations of Science, 4*, 537–542.
Hacking, I. (1997a). Repression and dissociation- A comment on "Memory repression and recovery". *Health Care Analysis, 5*, 112–135.
Hacking, I. (1997b). An Aristotelian glance at race and the mind. *Ethos, 25*(1), 107–112.
Hacking, I. (1998). Canguilhem amid the cyborgs. *Economy and Society, 27*(2–3), 202–216.
Hacking, I. (1999). The time frame problem: The law, social construction, and the sciences. *The Social Science Journal, 36*(4), 563–573.
Hacking, I. (2000a). How inevitable are the results of successful science? *Philosophy of Science, 67*, 58–71.
Hacking, I. (2000b). What mathematics has done to some and only some philosophers. *Proceedings of the British Academy, 103*(p), 83–138.
Hacking, I. (2001a). Dreams in place. *The Journal of Aesthetics and Art Criticism, 59*(3), 245–260.
Hacking, I. (2001b). La qualité. In J. Benoist, J. Bouveresse, & I. Hacking. *Quelle philosophie pour le XXIe siècle? L'Organon du nouveau siècle* (pp. 105–151). Paris: Gallimard.
Hacking, I. (2001c). Aristotelian categories and cognitive domains. *Synthese, 126*, 473–515.
Hacking, I. (2001d). *Les Classifications naturelles*. Cours au Collège de France. http://www.ianhacking.com/collegedefrance.html
Hacking, I. (2001e). Degeneracy, criminal behavior, and looping. In D. Wasserman & R. Wachbrott (Eds.), *Genetics and criminal behavior* (pp. 141–167). Cambridge: Cambridge University.
Hacking, I. (2002c). Inaugural lecture: Chair of philosophy and history of scientific concepts at the Collège de France. *Economy and Society, 31*, 1–14.
Hacking, I. (2002e). 'Vrai', les valeurs et les sciences. *Actes de la recherche en sciences sociales, 141–142*, 13–20.
Hacking, I. (2002f). *Façonner les gens*. Cours au Collège de France. http://www.ianhacking.com/collegedefrance.html
Hacking, I. (2003). L'Importance de la classification chez le dernier Kuhn. *Archives de Philosophie, 66*, 389–402.
Hacking, I. (2004a). Between Michel Foucault and Erving Goffman: Between discourse in the abstract and face-to-face interaction. *Economy and Society, 33*(3), 277–302.
Hacking, I. (2004b). The race against time: The hot money is on 'brain science'. Why? Because the people who hold the purse strings are getting older and fear dementia. *New Statesman, 133*, 26–28.
Hacking, I. (2004c). *Le corps et l'âme au début du vingtième siècle*. Cours au Collège de France. http://www.ianhacking.com/collegedefrance.html
Hacking, I. (2004d). Review. Truth and truthfulness: An essay in genealogy by Bernard Williams. *Canadian Journal of Philosophy, 34*(1), 137–148.

Hacking, I. (2005b). Truthfulness. *Common Knowledge, 11*(1), 160–172.
Hacking, I. (2005c). *Rethinking interdisciplinarity*. www.interdisciplines.org/interdisciplinarity/papers/7
Hacking, I. (2005d). Why race still matters. *Daedalus, 134*, 102–116.
Hacking, I. (2005e). *Façonner les gens II*. Cours au Collège de France. http://www.ianhacking.com/collegedefrance.html
Hacking, I. (2006). Genetics, biosocial groups & the future of identity. *Daedalus, 135*, 81–95.
Hacking, I. (2007b). Trees of logic, trees of porphyry. In J. L. Heilbron (Ed.), *Advancements of learning. Essays in honour of Paolo Rossi* (pp. 219–261). Florence: Leo S. Olschki.
Hacking, I. (2007e). *Les philosophes des sciences et les secrets de la nature*. Discours donné à l'université de Cordoba (Argentine), 14 mars 2007. http://www.college-de-france.fr/media/ian-hacking/UPL9050543492301304667_Discours_Cordoba.pdf
Hacking, I. (2007f). Putnam's theory of natural kinds and their names is not the same as Kripke's. *Principia, 11*(1), 1–24.
Hacking, I. (2007g). The contingencies of ambiguity. *Analysis, 67*(4), 269–277.
Hacking, I. (2008a). Philosophy of experiment: Illustrations from the ultracold. In R.de A., Martins y otros (Eds.), *Filosofia e História da Ciência no Cone Sul. Seleçao de Trabalhos do 5° Encontro* (pp. 17–29). Florianópolis: AFHIC.
Hacking, I. (2008b). Deflections. In S. Cavell, C. Diamond, J. Mc. Dowell, I. Hacking, & C. Wolfe, *Philosophy & animal life* (pp. 139–172). New York: Columbia University.
Hacking, I. (2008c). The suicide weapon. *Critical Inquiry, 35*(1), 1–32.
Hacking, I. (2009b). Humans, aliens and autism. *Daedalus, 138*(3), 44–59.
Hacking, I. (2009d). La Mettrie's soul: Vertigo, fever, massacre, and *The natural history*. *Canadian Bulletin of Medical History, 26*(1), 179–202.
Hacking, I. (2009e). How we have been learning to talk about autism: A role for stories. In E. Kittay, & L. Carlson (2010), *Cognitive disability and its challenge to moral philosophy* (pp. 261–278). Malaysia: Wiley-Blackwell.
Hacking, I. (2009f). Autistic autobiography. *Philosophical Transactions: Biological Sciences, 364*(1522), 1467–1473.
Hacking, I. (2010b). Pathological withdrawal of refugee children seeking asylum in Sweden. *Studies in History and Philosophy of Biological and Biomedical Sciences, 41*, 309–317.
Hacking, I. (2010c). Lloyd, Daston, nurture, and style. *Interdisciplinary Sciences Reviews. History & Human Nature Issue, 35*(3–4), 231–240.
Hacking, I. (2010d, April 27). *Lecture III-B. Historical derivation*. [Unpublished]. México: UNAM.
Hacking, I. (2011). Prueba, verdad, manos y mente. *Cuadernos de Epistemología,* 5, Universidad del Cauca, pp. 11–37.
Hacking, I. (2012a). Objectivity in historical perspective. Four comments on Daston & Galison. *Metascience, 21*, 11–39.
Hacking, I. (2012b). 'Language, truth and reason'. Thirty years later. *Studies in History and Philosophy of Science, 43*(4), 599–609.
Hacking, I. (2013a). What logic did to rhetoric. *Journal of Cognition and Culture, 13*, 419–436.
Hacking, I. (2013b, May 13). *Making up autism*. Inaugural C. L. Oakley Lecture in Medicine and the Arts, University of Leeds.
Hacking, I. (2015a). Let's not talk about objectivity. In F. Padovani et al. (Eds.), *Objetivity in science* (Boston studies in the philosophy and history of science, 310) (pp. 19–33).
Hacking, I. (2015b). Probable reasoning and its novelties. In T. Arabatzis, J. Renn, & A. Simões (Eds.), *Relocating the history of science. Essays in honour of Kostas Gavroglu* (Boston studies in philosophy of science, 312) (pp. 177–192).
Hacking, I., & Kirsch, M. (2003). Para-Marx et "le monde (des sciences)", *À quoi sert la philosophie des sciences?* 41, Rue Descartes, pp. 82–95.
Jardine, N. (2000). *The scenes of inquiry*. Oxford: Claredon.
Lynch, R. (2011). Foucault's theory of power. In D. Taylor (Ed.), *Michel Foucault: Key concepts* (pp. 13–26). London: Routledge.

Martínez, M. L. (1997). Paradigmas y Estilos de Razonamiento ¿Metaconceptos alternativos? In M. Otero (Org.), *KUHN hoy* (pp. 59–83). Montevideo: Facultad de Humanidades y Ciencias de la Educación.
Martínez, M. L. (2005). El realismo científico de Ian Hacking: de los electrones a las enfermedades mentales transitorias. *Redes: revista de estudios sociales de la ciencia, 11*(22), 153–176.
Martínez, M. L. (2007). Nuevos aportes de Ian Hacking a la historia y la filosofía de la ciencia. In P. Lorenzano, y H. Miguel (Eds.), *Filosofía e Historia de la Ciencia en el Cono Sur*. (Vol. 2, pp. 329–336). Buenos Aires: AFHIC-C.C.C. Educando.
Martínez, M. L. (2009a). Ian Hacking's proposal for the distinction between natural and social sciences. *Philosophy of the Social Sciences, 39*(2), 212–234.
Martínez, M. L. (2009b). Nominalismo y clases en ciencias humanas. *Galileo*, 40, 2da época, pp. 41–64.
Martínez, M. L. (2010). Ontología histórica y nominalismo dinámico. La propuesta de Ian Hacking para las ciencias humanas. *Cinta de Moebio. Revista de Epistemología de Ciencias Sociales, 39*, 130–141.
Martínez, M. L. (2013). El papel de la representación en la ciencia según Ian Hacking. In M. Martini (Ed.), *Dilemas de la ciencia. Perspectivas metacientíficas contemporáneas* (pp. 39–61). Buenos Aires: Biblos.
Martínez, M. L. (2016a). Aportes para un lenguaje de la coproducción. In M. Martini, y R. Marafioti (Eds.) (2016), *Pasajes y paisajes. Reflexiones sobre la práctica científica* (pp. 171–194). Moreno: UNM Editora.
Martínez, M. L. (2016b). Foucauldian imprints in the early works of Ian Hacking. *International Studies in the Philosophy of Science, 30*(1), 69–84.
Miller, J. (2011). *La pasión de Michel Foucault*. Santiago de Chile: Tajamar Editores.
Morey, M. (2014). *Escritos sobre Foucault*. Madrid: Sexto Piso.
Otero, M. (2004). Sobre las presuposiciones de la ciencia: el Essay on Metaphysics (1940) de Robin Collingwood como antecedente de toda una época. *Llull, 27*, 117–129.
Peters, M. A. (2007). Kinds of thinking, styles of reasoning. *Educational Philosophy and Theory, 39*, 350–363.
Rabinow, P. (Ed.). (1984). *The Foucault reader*. New York: Pantheon Books.
Revel, J. (2014). *Foucault, un pensamiento de lo discontinuo*. Buenos Aires: Amorrortu.
Ritchie, J. (2012). Styles for philosophers of science. *Studies in History and Philosophy of Science, 43*, 649–656.
Roudinesco, E., Canguilhem, G., & Derrida, J. (1996). *Pensar la locura. Ensayos sobre Michel Foucault*. Buenos Aires: Paidós.
Russell, B. (1905). On denoting. *Mind, 14*, 479–493.
Sciortino, L. (2016). Styles of reasoning, human forms of life, and relativism. *International Studies in the Philosophy of Science, 30*(2), 165–184.
Taylor, D. (Ed.). (2011). *Michel Foucault: Key concepts*. London: Routledge.
Tsou, J. (2007). Hacking on the looping effects of psychiatric classifications: What is an interactive and indifferent kind? *International Studies in the Philosophy of Science, 21*(3), 329–344.
Turner, E. (1988). Gravitational lenses. *Scientific American, 259*, 54–60.
van Fraassen, B. (2006). Representation: The problem for structuralism. *Philosophy of Science, 73*, 536–547.
Veyne, P. (1984). *Cómo se escribe la historia. Foucault revoluciona la historia*. Madrid: Alianza.
Wartenberg, T. E. (1984). Foucault's archaeological method: A response to Hacking and Rorty. *The Philosophical Forum, 15*(4), 345–364.

Index

M. L. Martínez Rodríguez, *Texture in the Work of Ian Hacking*, Synthese Library 435, https://doi.org/10.1007/978-3-030-64785-8

Printed in the United States
by Baker & Taylor Publisher Services

Printed in the United States
by Baker & Taylor Publisher Services